The Making of a New Science

Giorgio Ausiello

The Making of a New Science

A Personal Journey Through the Early Years
of Theoretical Computer Science

 Springer

Giorgio Ausiello
Dipto di Ingegneria Informatica,
Automatica e Gestionale
Università di Roma "La Sapienza"
Rome, Italy

ISBN 978-3-030-09679-3   ISBN 978-3-319-62680-2   (eBook)
https://doi.org/10.1007/978-3-319-62680-2

This Springer imprint is published by the registered company Springer Nature Switzerland AG
The registered company address is: Gewerbestrasse 11, 6330 Cham, Switzerland

In memory of my mentor
    Corrado Böhm,
    curious and creative researcher,
    pioneer of the theory of programming languages

In memory of Maurice Nivat,
    scientist,
    friend,
    founding father of European theoretical computer science

The sciences do not try to explain, they hardly even try to interpret, they mainly make models.

John von Neumann

The best practice is inspired by theory.

Donald Knuth

Preface

It was the year 1963 when I first came in contact with the world of electronic computing. Until then, in my high school years and afterwards as a freshman at university I had only heard about "electronic brains", *cervelli elettronici*, as they were called in Italy in the 1950s. The first computers had been installed in Italy in 1954, and in 1963 only 510 computers were operating in my country, essentially in universities, public administration, banks and large companies. The first "high level" programming languages had been created only in the mid-1950s and programming courses had just started being taught at university.

But aside from the new technological creatures that would later take up so much space in our modern life, in academia, computer scientists were trying to formalize and define in mathematical terms the concepts of computation, programming language, semantics of programs, and complexity of algorithms, and to understand their properties in order to improve the use of computers, the correctness of their functioning, the efficiency of the execution of programs. In other words they were laying the foundations of a new scientific domain: the theory of computing. Although its roots were nourished by mathematics, the new science was expanding beyond mathematics, its representation and modeling power had to cover not only the physical realm but all aspects of a human artifact, the world of computers and of computing: machines, algorithms, languages, programs, data, etc. The results achieved in the various strands of such theory were essential in the design of new computers, new programming languages, new computing environments, new applications. Beyond this, the theory of computing started to reveal a strong epistemological power: the need to automatize functions of human life (in science, industry, commerce, banking, public administration, arts, games, etc.) required the ability to abstract concepts, to create models, to understand and describe processes, to classify and organize data.

While taking my first steps in the field of computer science I had the chance to witness from the beginning the development of such ideas, to meet some of the main figures in this world, to breathe the exciting atmosphere that circulated in universities and research centers. To phrase it with the words used by Richard Karp in 1985 when he received the Turing Award: "there was a very special spirit in the air; we knew that we were witnessing the birth of a new scientific discipline centered on the computer". When in the 1970s the first European society for the promotion of this new science was created and the first European scientific journal devoted to it was founded, the name was already well established: theoretical computer science.

In these pages I have provided the readers with my view of the birth of this discipline and of the events that marked its development in the early days. Through my personal memory, tracing back the concepts I learned, the people I met, I have tried to present the ideas that were taking their shape in those years and the roads undertaken by the first explorers. Of course it is a partial view, based on my personal experience. My perspective starts from the fourth floor of the building of the Italian National Research Council where the Istituto per le Applicazioni del Calcolo was located in the 1960s and where I met for the first time a computer. From there the view extends to the Italian scene in the early 1970s and then to the European scale to report the early efforts to build a European theoretical computer science community. Following schools, conferences and publications in which the new concepts were taking shape I have tried to give an account of the making of the new science, of the formation of some of the main chapters of theory of computing. As I said it is a partial and biased view, both in terms of the subjects whose developments I have been able to describe (strongly related to my fields of interest), and in terms of the geographic communities whose activity I had the opportunity to interact with. I want to apologize for not being able to mention all the people and all the achievements that have paved the way to the birth and growth of theoretical computer science.

The history that I have tried to trace starts in the early 1960s and ends in the late 1980s, through about twenty five years. The beginning is related to the moment at which, as a physics student, in 1963, I entered the class room of Corrado Böhm and I got in touch with the universe of computer science and of its mathematical foundations. The end of the history is more fuzzy. By the late 1980s the theory of computing had become a mature scientific field, whose principles had become course material in all universities through-out the world. Besides, in the same period, computer systems and computer applications were starting to undergo a deep transformation, related to tech-nological developments and to the advent of the Internet and of the Web, a transformation that would definitively change the world of computing and substantially broaden its foundations.

Our young colleagues and students, who now work on well-defined tracks, in a field that has been systematized through the effort of generations of researchers, but that is also dramatically changing under the pull of a re-lentless technological drive, are often not aware of how their research field developed in the early days. My hope is that they can profit by learning how the theory of computing started to take shape. Also young scientists and professionals working in other disciplines, curious to know what is hidden behind the screen of a computer, reading these notes may understand how our knowledge about the principles of computing has been gradually formed. To allow them to understand these notes I have tried to make most concepts simple and to add here and there some tutorial remarks.

A special thank goes to my friend Giorgio Gambosi, professor of theoretical computer science at the University of Rome "Tor Vergata", who revised and edited these notes with competence and patience and who provided valuable technical contributions to make the volume accessible also to non specialists. I also wish to thank Aldo de Luca, Josep Díaz, Jozef Gruska, Titti Guerra, Pino Italiano, Fabrizio Luccio, Ugo Montanari, Mike Paterson, Antonio Restivo, Domenico Saccà, and Andrzej Salwicki for having provided pictures and/or personal memories and archive material and, in some cases, for having improved my recollection of facts and events. These notes could have not been written without the support of Internet resources. The huge collection of major (or minor) information contained in Wikipedia and in a variety of other repositories has been helpful to sustain my memory in recovering facts and figures of the theoretical computer science world from 40 or 50 years ago. Still I alone am responsible for mistakes and omissions.

March 2018   Giorgio Ausiello
                                                     'Sapienza' University of Rome

Contents

Chapter 1
Tubes

The computer room at the Istituto Nazionale per le Applicazioni del Calcolo (INAC) was 200 square meters large, on the 4th floor of the building of the Italian National Research Council, a rationalist building rising on what at that time was Piazzale delle Scienze (now Piazzale Aldo Moro) in Rome. The computer was a Mark I machine, built by the English company Ferranti. The machine was also known as Manchester Universal Electronic Computer DC.4 because it had indeed been built on the basis of a prototype designed at Manchester University.

The machine had been bought by the Italian government upon request of the great mathematician Mauro Picone[1], founder of INAC in 1932, and Director of the Institute for almost thirty years. It was the second computer installed in Italy, the first one having been installed at Milan Polytechnic in 1954. It is known that two INAC researchers, Enzo Aparo and Corrado Böhm, had gone to Manchester in order to test its computation power[2].

The installation of the Ferranti machine had taken place in 1955 and had been done by Giorgio Sacerdoti, one of the first Italian engineers with specific skills in electronic computers, who later was hired by Olivetti and would become one of the leaders in their effort to develop electronic computers. The machine was inaugurated on December 14, 1955 and was nicknamed "FINAC" (Ferranti INAC). Since July 1956 the Director of the computing center was Paolo Ercoli, who also contributed to improving various technical characteristics of the machine. Words were 20 bit long and one page consisted of 64 words whose bits were displayed on the surface of a Williams

---

[1] Actually, before deciding to buy the Ferranti machine, Mauro Picone made several (but unsuccessful) attempts to convince the National Bureau of Standards and the Stanford Research Institute to cooperate with INAC in the construction of a new machine that might have been built in Italy (see Pietro Nastasi, *Picone, il calcolo automatico e FINAC, una storia lunga 30 anni*. In *50 anni di informatica in Italia* PRISTEM/Storia 12–13, Università Bocconi, 2005).

[2] The test had been very satisfactory. It had consisted in the solution of 62 linear equations in as many variables and had required 'only' 3 hours and 15 minutes.

© Springer International Publishing AG, part of Springer Nature 2018
G. Ausiello, *The Making of a New Science*,
https://doi.org/10.1007/978-3-319-62680-2_1

Fig. 1.1 The FINAC computer

cathode ray tube in a 64×20 matrix of tiny luminescent dots. The machine was equipped with 4000 vacuum tubes, 9 km of cables, a secondary storage consisting of a magnetic drum with a capacity of 2^{15} words and punched tape and teletype as input/output devices (see Fig. 1.1). Two long cabinets contained all the circuitry. In the center was the console, with hundreds of switches and buttons, and on its side stood the cathode ray tube.

It was Spring 1964 when I went for the first time into the room. At that time I was an undergraduate student at Rome University (now Sapienza University of Rome but then there was only one public university in Rome). As a third year student in physics (identification number 32534) in Fall 1963 I had to choose my curriculum. Most of my colleagues were attracted by the fascinating studies of nuclear and particle physics carried out by Edoardo Amaldi and his internationally renowned group of scientists but I rather decided to follow the electronics curriculum.

During the summer of 1963 I had read a book (*Il pensiero artificiale*, by Pierre de Latil) explaining in simple terms the principles of control systems and the regulatory effect of feedback in natural and artificial systems. I had also seen some tutorial books concerning digital circuits. But I think that what was more interesting to my eyes in digital systems was the logic underlying the physical devices. Although I was still unaware of the formal techniques of circuit synthesis, in my first electronic laboratory I felt excited to design a circuit consisting of AND, OR, NOT gates, starting from the Boolean function that I was supposed to realize. In any case the electronics curriculum in the degree in physics did not have anything to do with computers: it was oriented toward the design of electronic instruments needed for nuclear physics experiments. In the electronics lab the task assigned to my group was to realize a particle detector and counter. It was interesting and I took part in the project with great enthusiasm. One day a colleague called me on the phone: 'Have you heard about the new course that Professor Böhm is giving for students in mathematics and in physics? It is a course on programming techniques (the title indeed was *Tecnica della programmazione*) but it also addresses topics in cybernetics like automata theory!'. We did not really have an idea regarding what the course was about, neither what cybernetics or automata theory were, but my intuition told me that this was what I was looking for.

If you had known Corrado Böhm you might easily understand that the course did not have much to do with what today we would consider a computer programming course. Neither was it related with what I later understood to be cybernetics. Indeed it was a course in what we would now call 'Theory of computing'. The main chapters of the course were devoted to the abstract computational models that form the logical foundation of computer programming: Turing machines, λ-calculus, combinatory logic, Markov algorithms. The only part of the course really devoted to programming was the one in which Corrado taught an interpretative language that he had designed for the FINAC machine, called INT-INT. The acronym meant a combination of interpretation and integration, but despite its applied flavor I remember that in my homework assignments I enjoyed much more to implement Turing machines than to write INT-INT programs.

Actually in that period I also took a real computer programming course. It was given in the Institute of Physics but it was not an official course. It was a short course in which the lecturer used to teach FORTRAN (one of the first programming languages, especially designed for numerical applications) to students and professors who needed to use the computer for their scientific work. But the impact that Corrado Böhm's lectures had on me was totally different.

In fact listening to Corrado's lectures on Turing machines and on the other computation models was one of the events that changed my life. Sitting in the classroom I discovered what no mathematics course had previously taught me, I discovered the fascinating world of the logical foundations of computer science, the mathematical studies that had prepared the road for the construction of the first electronic computers.

For one of those rare events in the history of thinking, magically in the 1930s (only thirty years before, for me, at that time) in a Europe already in anguish due to the totalitarian regimes dominating the continent, some of the most extraordinary mathematical minds had given rise to a new era of mathematical logic, the era of computability. Indeed less than thirty years had passed since Kurt Gödel and Alan Turing in Europe and other logicians such as Emil Post, Alonzo Church, and Stephen Cole Kleene in the US had established the formal definition of the concept of computability and had discovered the existence of functions that could not be computed in algorithmic terms and problems that could not be solved by means of automatic computations.

A special role in such discoveries was played by the British mathematician Alan Mathison Turing. In the Spring of 1935, while he was still a 22-years-old student in Cambridge, after listening to a lecture concerning the incompleteness of the logical theory of arithmetic, proved by Gödel a few years before, Turing decided to work to a closely related problem, the decision problem for arithmetic[3]. Such a problem required to establish whether there exists an algorithm to decide if a logical statement expressed inside the theory of arithmetic is a theorem, i.e. it can be derived from the axioms, or not.

Alan Turing (Fig. 1.2) solved the problem showing that such an algorithm does not exist, but in order to achieve his result Turing made three steps of the greatest relevance for the theory of computing. First he presented an (abstract) computing device (later called 'Turing machine') and provided convincing arguments in support of the idea that any algorithmic computing process could be executed by such a computing device; second he showed that there exist Turing machines (called 'universal Turing machines') that are able to mimic any other Turing machine; finally he proved that there are problems that none of his devices (and hence no computing device of any nature) is able to solve, among them the decision problem for arithmetic.

Fig. 1.2 Alan M. Turing

The impact of Turing's result has been far reaching[4]. His paper *On Computable Numbers with an Application to the Entscheidungsproblem*, published in 1936, several years before the first real electronic computers were conceived, in its own right can be considered the beginning of theoretical computer science for the mark it made on the history of computing. Man had used algo-

---

[3] The decision problem (*Entscheidungsproblem*) for the logical theory of arithmetic was formulated by the German mathematician David Hilbert (1862–1943) at the Congress of the International Mathematical Union in Bologna in 1928, together with the related problems regarding the consistency and the completeness of arithmetic. David Hilbert is well known for having presented a list of 9 open problems of mathematics (that became 23 in the subsequent written version of the paper) in a famous conference held at the Congress of Mathematicians in Paris, in 1900. The consistency of arithmetic, that is the fact that in the logical theory of arithmetic it is not possible to prove two contradictory statements, was one of these problems. In 1931 Kurt Gödel proved that arithmetic is either incomplete or inconsistent.

[4] The role of Alan Turing (1912–1954) in computer science is well represented by the fact that the most prestigious award in the field (sometimes referred to as the Nobel Prize of Computing) derives its name from Alan Turing. The 'Turing Award' has been assigned every year since 1966 by the Association for Computing Machinery (ACM).

rithms for thousands of years but only now the power of algorithms and, at the same time, their limitations were being understood, and proved in mathematical terms. Only now mankind could realize that not all mathematical problems could be solved by means of algorithms.

All these aspects were presented to us in Böhm's lectures. Thirty years later the results of Turing and of the other logicians appeared in all their powerful consequences. Turing's big achievement was not only relevant for mathematics (in a memorable talk given at Rome University in the early 1970s, the great Russian mathematician Kolmogorov declared that Turing's results concerning computability were the most important achievements of mathematics in the 20th century) but had an immediate impact on electronic computers. First of all the existence of universal machines (that is machines able to interpret the description of any Turing machine and to mimic its behavior) was at the basis of the notion of a computer program and of the development of programmable computers. In second place the limitation of the power of algorithms could be seen as a limitation of the power of computers, a surprising result since in popular imagination they were still considered almighty machines.

In the lectures Corrado Böhm presented to us also the other computation models studied by mathematicians in the 1930s, different in terms of computing paradigms but equivalent in terms of computation power, Church's λ-calculus, Curry's combinatory logic, Post's rewriting systems, and he explained the importance of the so-called Church's Thesis (or Church-Turing Thesis), the conjecture (still not disproved) that the computation power of Turing machines indeed captures any possible notion of an algorithmic process. Through his lectures we also realized how, in the 1950s and in the early 1960s, building on the basis of Turing's Thesis, the various computing paradigms started to influence the design of the first programming languages. The course was integrated with some lectures given by a young pupil of Corrado, Salvatore Caporaso, on another fascinating topic: Markov algorithms, a computation model based on rewriting rules that had been invented by the Russian mathematician in the 1940s.

As students, through Corrado's lectures, we could perceive that a new science was being born, arising from the roots of mathematical logic and projecting its light on the future of computers and of computer programming. Corrado used to convey an incredible enthusiasm for the study of computation processes and especially for the interplay between languages and computation. Since Wolf Gross, a Polish mathematician hired by CNR after the war[5], introduced him to λ-calculus and combinatory logic, Corrado's scientific interest was concentrated on algorithmic languages, that is languages in which the syntax provides the description of an algorithm and at the same time

---

[5] Wolf Gross (1920–1991) was born in Bielski (Poland). He fled to Rome during the war and was hired by CNR in the 1950s. I met him in 1963 as he was Professor of 'Mathematical physics' at Rome University, a beautiful course in which all what we had learned in the courses of general physics was revisited in terms of abstract mathematical modeling.

the machinery needed to carry on the computation process. Such intellectual passion used to flow in Corrado's lectures. Collaborating with him Giuseppe (Puccio) Jacopini[6] and Andreina Morelli contributed to enrich the students' curiosity with examples and exercises, partly to be done on paper, partly to be run on the FINAC machine.

This is how I came in contact for the first time with a computer. FINAC had been equipped by Puccio Jacopini with a simulator of Turing machines. One had to design a Turing machine on paper and then feed it as input to the simulator. The input to FINAC had to be provided on a punched tape. The output was printed on paper by a teletype. When an error occurred, debugging required us to cut the tape with a pair of scissors and replace the incorrect part with a new part to be attached to the old one by means of scotch-tape. Our first homework was to design and execute Turing machines performing arithmetical operations with decimal numbers. I did my assignment at home and then I had to encode my machine on the punched tape and then run it on the computer.

Entering the computer room was allowed only with special permission. Only a few people could sit in front of the console and interpret the blinking of the cathode ray tube to check whether the computation was being performed correctly. The machine also produced a sound. Any program used to produce its own sound through a loudspeaker and the sound was helpful to understand how the machine was behaving, what parts of the circuitry were involved at a given point of the computation. Sound was even useful for debugging. When a monotone continuous sound was perceived it meant that the program execution was stuck in a never-ending cycle.

Using the computer was a privilege for us. Many INAC researchers were still making their calculations on big black Brunsviga mechanical calculators that had been bought in the 1950s and that equipped all rooms of the Institute. Only part of the mathematical work of INAC was processed on the computer, notably the FINAC machine had been used for the engineering calculations for the construction of the big Vajont dam around 1955.

In June 1964 I passed the exam of *Programming techniques* with a score of 30 "cum laude" and announced to Corrado Böhm that I would come to visit him again to ask him about the subject for my 'Laurea' (Master's) thesis.

Corrado Böhm was born in Milan in 1923[7]. He lived there until 1942 when he moved to Lausanne where he graduated in Electrical Enginering in 1946. He then was enrolled at ETH in Zurich where he took his PhD degree in

---

[6] Giuseppe Jacopini (1936–2001) was one of the most brilliant CNR researchers in mathematics and theoretical computer science. He was always reluctant to write papers but his scientific production (although limited) shows how deep and original were his intuitions about the forms and the power of computing. His collaboration with Corrado Böhm gave excellent results and it is in particular tied to the Böhm-Jacopini Theorem, a result that stands at the basis of the methods and languages of 'structured programming'.

[7] A beautiful account of the first years in the scientific life of Corrado Böhm can be found in A. Restivo, *Gli esordi scientifici*. In *L'informatica. Lo sviluppo economico, tecnologico e scientifico in Italia*, edited by F. Luccio, Edifir, 2007.

Mathematics in 1952. Formally, his supervisor was Paul Stiefel but indeed it was the great Swiss mathematician Paul Isaac Bernays (who, according to Corrado's words, "taught him about Turing machines and Turing's work on computability") who inspired Corrado's work. In Zurich Corrado had the chance to work on a Zuse computing machine Z4 and started to conceive the idea of symbolic programming languages and of automatic translation from high-level languages to machine language.

His thesis was devoted to one of the first scientific contributions in the field of programming languages: the design of a programming language and of a compiler[8] written in "source" language, that is the same high-level language that the compiler itself was supposed to translate[9], an example of what would now be called 'meta-circular compiler'. Besides, as acknowledged by D. E. Knuth and L. T. Pardo in their paper *The Early Development of Programming Languages*, "the language was interesting in itself because every statement (including input statements, output statements and control statements) was a special case of an assignment statement". They also underline that "of all the authors we shall consider, Böhm was the only one who gave an argument that his language was universal, namely, capable of computing any computable function". Another important contribution (for which he filed a patent in Italy in 1952) was the design of a computer able to perform symbolic computations.

After working for a short period at Olivetti in Ivrea, in 1953 Corrado was hired at the National Research Council and started to work at INAC (cf. Fig. 1.3). His research work throughout those years was essentially devoted

Fig. 1.3 Corrado Böhm and his colleagues at INAC in the late 1950s (from left to right: Corrado Böhm, Paolo Ercoli, Wolf Gross, Aldo Ghizzetti, the Director Mauro Picone, n.i., Roberto Vacca, Enzo Aparo, Puccio Jacopini)

---

[8] A compiler is a program whose task is to translate a program written in a high-level programming language (source code) to a program written in machine language (object code) that can be directly executed by the computer.

[9] C. Böhm, *Calculatrices digitales. Du déchiffrage des formules mathématiques par la machine même dans la conception du programme*, Annali di Mat. Pura e Applicata, series IV, vol. XXXVII, 1–51, 1954.

to numerical calculus. Subsequently he moved his interest toward studying connections between the computation models that had been defined in the 1930s with the aim of characterizing the concept of algorithm and the class of computable functions (Turing machines and λ-calculus, in the first place) and computer programming languages that were just being introduced in the late 1950s.

As I understood later on, the basic idea underlying most of Corrado's work was to turn such abstract computation models into programming languages and in such way to find out what were the intrinsic structural properties of programs. The most remarkable example of this research thread was the famous result known as the "Böhm-Jacopini Theorem" that essentially states that all algorithms can be designed by making use of only three fundamental programming constructs: sequencing (that requires the execution of instructions one after the other, with no jumps), selection (if a certain property is true then execute the subprogram $p1$ else execute the subprogram $p2$) and iteration (while a certain property holds do subprogram p).

This result, which appeared in the paper *Flow Diagrams, Turing Machines and Languages with Only Two Formation Rules* (that has now reached 900 citations), has been the basic foundational support of the approach to programming known as "structured programming". Ten years later structured programming constructs were at the core of any programming language. Currently, all the most common languages used in programming (such as Java, C++, C#, Python, and so on) are based on the fundamental structures studied by Böhm and Jacopini.

One of the guidelines that inspired Corrado's work was the idea that a program could be written as a sequence of symbols, just as a mathematical formula. Despite the title of his most famous paper, he hated flow charts (he never mentioned them during his lectures) and was convinced that conditional and unconditional jumps in programs would disappear from future programming languages (as indeed happened). In the early 1960s Corrado had been one of the first scientists in the world to study λ-calculus and to foresee that it might be a paradigm of what a functional programming language would be in the future. In 1963 he had been the advisor of the Master's Thesis in Mathematics of Marisa Venturini Zilli in which it was shown how the data structures of Iverson's language APL (a language introduced in 1962 and based on a mathematical notation for operations on multidimensional arrays previously introduced by Iverson in 1957) could have been represented in λ-calculus.

Working on λ-calculus and combinatory logic, together with Wolf Gross, he had developed a new algorithmic language, called CUCH (from the first letters of the names of the two logicians Haskell B. Curry[10] and Alonzo

---

[10] Haskell B. Curry (1900–1982) was an American logician. He obtained his PhD in Göttingen under the supervision of David Hilbert. His work on combinatory logic (developed on the basis of ideas of Moses Schönfinkel with the aim to provide a foundational ground to mathematics) had a strong conceptual impact on the definition of functional

Fig. 1.4 J. Barkley Rosser (left), Haskell B. Curry (center), Alonzo Church (right) in 1982

Church[11] (Fig. 1.4)) and part of his efforts were devoted to show how powerful CUCH was for describing computational structures.

Let us see how he described his research efforts at that time[12]. "In that period I was interested to find the 'best' computation model among Turing machines, Post production systems, Markov algorithms, Curry's combinatory logic (CL) and Church's lambda-calculus (LC). The fundamental idea was to transform the 'winning' model into a powerful and synthetic programming language. The paper that is at the base of structured programming was born from the examination of the structures of some Universal Turing Machines, together with Jacopini's intuition that the necessary restrictions might have been represented in terms of flow diagrams. At the same time I was convinced that indeed with CUCH, a fusion of CL and LC, I had found the best abstract machine".

---

programming languages. Two well-known programming languages (Haskell and Curry) are named after him. Also the operation of transforming functions with multiple arguments into a sequence of one-argument functions takes the name of 'currying'.

[11] Alonzo Church (1903–1995) was an American logician who gave outstanding contributions to the mathematical foundations of computer science. His name is related to the definition and development of λ-calculus, whose computational power was demonstrated by Church and Turing to be equivalent to the power of Turing machines. This result gave rise to the well-known Church-Turing Thesis which, as we said already, states that all computable functions can be computed by means of Turing machines (or, equivalently, by λ-calculus computations). As combinatory logic, also λ-calculus had a tremendous impact in the definition of functional programming languages.

[12] C. Böhm, *Gleaning the Past and the Future in Computer Science*, Bulletin of EATCS, 74, 178–189, 2001.

In 1964 Raffaele Giovannucci, another student of Corrado, had written two reports inspired by the 1958 paper of Frederic Fitch *Representation of Sequential Circuits in Combinatory Logic*: in one it was shown how to represent combinatorial circuits by means of CUCH formulae, and in the second it was shown how to represent analog circuits. When I asked Corrado for a subject for my thesis he proposed to me to make another step in the same direction, that is to provide a CUCH representation of "hybrid (analog-digital) computers".

In the early 1960s analog computers were still used in several applications, mainly for the solution of systems of differential and integral equations (for example analog computers were widely used in the control of power plants and of industrial processes). In an analog computer a problem to be solved is modeled in terms of physical quantities associated to the computer components (usually electronic circuits), with the relations between input data and results implemented by a suitable use of the components themselves. In an analog computer, data are then represented as values of some measurable physical quantity in the system, while, as is well known, in a digital computer they are represented by a suitable coding of their numeric values.

This type of processing machine had been used in various forms for centuries (when mechanical gears were employed in place of electronic circuits) but reached their apex in the 1940s and in the 1950s when the full power of electronics could provide fast and precise solutions to various kinds of analytic problems. The advantage of analog computers with respect to digital computers, at that time, was both in the speed and in the precision (a limit that only two decades later would have been definitely overcome in favor of digital computers) while the main disadvantage was in the fact that analog computers were in a sense special purpose machines: the circuits were configured in such a way to solve a certain problem but when a new problem had to be addressed (or simply new equations had to be analyzed) the machine had to be suitably re-configured, for example by changing the connections over some switchboard. To obtain an improvement with respect to this aspect analog computers started to be equipped with digital circuits able to implement such re-configuration efficiently, giving rise to the new generation of so-called "hybrid computers".

CUCH had various characteristics that could prove useful in the description of hybrid analog-digital computers. Being a typeless language, variables could be used to represent any type of data (real numbers, real functions, logical functions assuming only TRUE, FALSE values, etc.). In second place CUCH operators could be used to model both integral or differential operators and logical gates.

I started to dig into the literature concerning analog and hybrid computers. I still remember reading dozens of technical reports of the Italian nuclear research center CNEN (where in particular Agostino Mathis was the guru of analog computing) and of the EURATOM research center located at Ispra, on Lake Maggiore, where some of the greatest experts in analog computing were working at that time. But the most difficult part of my work was to relate the abstract, far-reaching intuition of Corrado with the engineering low-level details of the technical literature. One day Corrado came to my place. From his bag he extracted a strip of paper that he rolled as a Moebius strip in front of my eyes (and in front of the astonished eyes of my mother who was trying to realize what kind of profession her son was going to start). As I understood later, his intention was to introduce a physical notion of "recursion" and to suggest a possible way toward "universality" in hybrid computers.

In any case, in the summer of 1965 I finished the exams of the physics curriculum (it is a pleasure now to remember how much I enjoyed the course of 'Theoretical physics' that was held by Marcello Cini) and I started working full time on my Thesis. It was precisely in that summer that I came across another stream of new ideas and mathematical concepts that were laying the foundations of the new science: formal language theory.

In the month of September 1965 a NATO School on Programming Languages (one of the first events of that kind in Italy) was held in Pisa. The school was part of the *Advanced Study Institutes Program* organized by the NATO Science Committee.

An analogous school had been held, the year before, in Ravello, on the coast of Amalfi's peninsula, organized by Eduardo Caianiello, professor of theoretical physics in Naples. Caianiello was the leader of important projects in cybernetics and the school, held in the wonderful scenario of Villa Cimbrone, had been a famous event and has remained a legend in the theoretical computer science community. Although the title, *Automata Theory*, was rather specific, addressing only a particular domain, the school had a very strong interdisciplinary character. The lecturers were outstanding, coming from a wide variety of fields: graph theorists such as Claude Berge, mathematicians like Marcel-Paul Schützenberger and Maurice Nivat, whose names we shall meet several times in these notes, linguists such as Maurice Gross, logicians like Martin Davis and Michael Arbib, experts in automata theory like Michael Rabin and neurophysiologists like Warren McCulloch. Also Corrado Böhm and Wolf Gross had taken part in the school (where they presented the language CUCH) and spoke to us about the relevance of such an event. The NATO School in Pisa (and, the subsequent year, the NATO School in Villard de Lans) was expected to have at least the same importance.

I applied as a student at the Pisa School and succeeded to be accepted with a full grant covering registration fee, food and lodging in a beautiful room in what was the "Scuola Normale Femminile", just in front of the river Arno. At that time Pisa was on its way to become the capital of computer science in Italy. One of the leading figures of the computer science school in that city was Alfonso Caracciolo[13] (Fig. 1.5), more precisely, Alfonso Caracciolo, Prince of Forino, Duke of Belcastro, heir of an ancient noble Neapolitan family. The

---

[13] Alfonso Caracciolo (1925–1996) was one of the founders of computer science in Italy, in particular the father of the theory of computing school in Pisa. As we will see he had

school was organized by François Genuys[14] (a French mathematician working with IBM) and Caracciolo was the local organizer and one of the lecturers. His lectures concerned the formal definition of a programming language for machine tools. Beside him, the board of lecturers consisted of some of the greatest computer scientists of that time. Among them Seymour Ginsburg, researcher at System Development Corporation (SDC), in Santa Monica, California, who lectured on context-free languages, Adriaan van Wijngaarden, Director of the Mathematisch Centrum in Amsterdam, who addressed the issue of metalanguages for formal languages, Peter Naur, one of the Danish pioneers of computer science, researcher at Regnecentralen, a Danish computer company, who spoke about the syntax of programming languages, and finally Peter Landin who spoke about programming languages inspired by λ-calculus.

Fig. 1.5 Corrado Böhm and Alfonso Caracciolo at the AICA Congress in 1968 (on the right side, Marisa Venturini, on the left side, Giorgio Sacerdoti, in the back Arrigo Frisiani, Giorgio Ausiello, Alfonso Miola

It may be interesting to spend a few words here to explain why these figures have a place in the history of our discipline. In order to do that we have to turn our attention to what was at that time one of the major research issues in computer science: the definition of the formal syntax of programming languages and the construction of compilers. Without mentioning the first experiments in high-level programming (such as the Plankalkül developed for the Zuse machine and the Autocode designed in Manchester in 1952 for the Mark 1 machine) we must realize that the first widely used general-purpose

---

an important role in the creation of the European Association for Theoretical Computer Science.

[14] François Genuys was famous for having computed 10,000 digits of π in 100 minutes on an IBM machine 704 for the first time.

programming languages were designed in the late 1950s. FORTRAN, whose name derived from the words "formula translation", a programming language suited for scientific computing, was designed at IBM by a group led by John Backus[15] between 1954 and 1956 and its compiler was delivered in 1957. Its counterpart for commercial applications, COBOL ('commercial business oriented language'), designed by CODASYL, a joint committee formed by universities, industry and US government, appeared in 1959.

The major software component of a compiler is the so-called parser that is supposed to identify the syntactic structure of a program and to produce the object code to be run on the target machine.

Due to the well-established theory of parsing resulting from those studies and to the related availability of tools which generate parsers from the description of the syntax of the corresponding language, the design of a parser is nowadays a school exercise for a third-year student in computer science with a good basis in formal languages and in compiler construction. In the 1950s the compilers of the first programming languages were defined in a rather ad hoc manner, without the guidance of methodological principles. Concerning this point, it is interesting to read what Edsger Dijkstra[16] says in the report *Algol 60 Translation*: "While the problem [to build a compiler] was yet new to us we began a few times by treating relatively simple tasks, but every solution we then found turned out later to be inadequate in more complicated cases". A major effort for computer scientists was therefore to design programming languages with a precise and clean syntax that could make easier the production of efficient and correct compilers. Again, theoretical studies and technological needs had the chance to meet each other.

In 1956 the American linguist Noam Chomsky had introduced a classification of languages based on generative tools: formal grammars. The idea is based on rewriting rules, a concept derived from formal logic, which define how a logic formula can be modified by substituting terms with different terms. In formal logic, the whole process of deduction from an axiom is described in terms of applying a sequence of suitable substitutions. In the Chomsky approach, a syntactic structure can be associated to each word in

---

[15] John Backus (1924–2007) obtained the Turing Award in 1977 with the following citation: "For profound, influential, and lasting contributions to the design of practical high-level programming systems, notably through his work on FORTRAN". Despite having received the award for his contribution to procedural languages, John Backus devoted a great part of his life to the design of functional languages. Emblematic is the title of his Turing lecture: *Can Programming Be Liberated from the von Neumann Style?*

[16] Edsger W. Dijkstra (1930–2002) is one of the giants of the early days of theoretical computer science. During his life he had an alternation of appointments in industry and in academia. In 1972 he received the Turing Award for his work related to the programming language ALGOL but also for the variety of contributions he had given in all fields of computer science, starting from his famous algorithm for computing shortest paths in a graph and including the theory of concurrent processes. All such contributions had an enormous practical impact on computer systems and computer applications. We will meet him again several times in these notes.

the language according to how that same word can be produced by a set of rewriting rules defined in the grammar and operating on a suitable collection of (non-terminal) symbols, starting from an initial symbol (the axiom). Defining constraints on how this rewriting can be done, by limiting the structure of rules, makes it possible to classify grammars and, correspondingly, languages generated by them.

At that time Chomsky (born in 1928) was affiliated with the Department of 'Modern Languages' and with the Research Laboratory of Electronics of MIT. The basic aim of his work was clearly stated in his paper: "We investigate several conceptions of linguistic structure to determine whether or not they can provide simple and 'revealing' grammars that generate all of the sentences of English and only these". The notion of grammar that he developed was inspired by the classical rules of English syntax: a 'sentence' consists of a 'noun phrase' followed by a 'verb phrase'; a 'verb phrase' consists of a 'verb' (in the intransitive case) or of a 'verb' followed by a 'noun phrase' (in the transitive case) etc. And were based on a rewriting mechanism that allows the replacement of a part of a string with other strings. His paper *Three Models for the Description of Language* played an important and unexpected role in the future development of computer science. In particular in his paper Chomsky introduced the 'finite-state grammars' and the 'phrase-structure grammars'. Each class of grammars gave rise to a corresponding class of languages that could be 'generated' by such grammars, in the sense that a language belongs to a class if it can be generated by a grammar of the corresponding class: 'finite-state languages' (the languages that we now call 'regular languages') and 'phrase-structure languages' (that are now known as 'context-free languages').

Pretty soon it appeared that Chomsky grammars, originally aimed at the study of natural languages, were too rigid to be applied to natural languages, while at the same time they emerged as the right instrument to characterize artificial languages and in particular programming languages. In particular the attention concentrated on context-free grammars. Regular grammars, in fact, do not allow the definition of nested structures, such as arithmetical expressions with nested parentheses, while context-free grammars, although still rather simple, allow nested structures and, more generally, allow the definition of syntactic concepts through repeated refinements: for example the concept of 'program' could be defined in terms of a 'sequence of statements', each 'statement' being (independently from the context) either an 'assignment statement' or a 'selective statement' or an 'iterative statement' and then any of these types of instruction could be further refined by specifying its syntactic structure.

When the first results concerning formal languages started to circulate, research in computer science rapidly developed in two directions: on one side toward the practical goal of applying context-free grammars to the syntactic definition of programming languages (and to the design of the corresponding compilers), on the other side toward the theoretical goal to define the best suited formalism to specify languages, that is to define the most appropriate 'meta-languages'.

Following the first line of research in the years 1958–1960 a group of researchers started to work on the design of a new programming language: Algol (whose name derived from the words "algorithmic language"). The major target in the design of Algol was to conjugate a clean syntax (the keywords "begin" and "end", widely applied to delimit code blocks, were introduced in the 1958 version of Algol) with new features like the syntactic scoping of variables, that is the specification of the part of a program where a variable is defined and can be referenced through the syntactic structure of the program itself.

From January 11th to January 16th, 1960, thirteen computer scientists (7 from Europe, among them Friedrich Bauer, Peter Naur and Adriaan van Wijngaarden, and 6 from the US, among them John Backus, John McCarthy and Alan Perlis) took part in a meeting in Paris that originated the final Algol 60 report. This was the second version of the language and had a tremendous impact on the design of several programming languages (Algol W, Pascal, Simula, Ada and the language C). It is interesting to note that, in the words of C.A.R. Hoare, who received the Turing Award in 1980 for "fundamental contributions to the design and development of programming languages", Algol was "a language so far ahead of its time that it was not only an improvement on its predecessors but also on nearly all its successors". The first implementations of Algol 60 were realized between 1960 and 1962. Apparently the very first implementation, operating since June 1960, was done by Edsger Dijkstra on an X1 machine, at the Mathematisch Centrum in Amsterdam[17].

The second line of research concerning the role of meta-languages for programming languages had a more foundational character. The point of view of many theoreticians was that before designing new programming languages and addressing practical issues concerning their implementation it was necessary to identify the suitable mathematical approaches to their formal description. An important forum of discussion of these ideas was the Working Group 2.2 (Formal Description of Programming Concepts) of IFIP[18] Technical Committee 2 (IFIP TC2) that had been created in 1964 by various scientists, among them Corrado Böhm, Peter Landin, Dana Scott, and Christopher Strachey.

---

[17] The Mathematisch Centrum in Amsterdam was founded in 1946. In 1997 the name was changed to Centrum Wiskunde & Informatica (CWI) to acknowledge the growing role that informatics played in the research activity of the center.

[18] IFIP, created in 1960 under the auspices of UNESCO, is the International Federation for Information Processing Societies, that is the world organization that gathers all national informatics societies (AICA in Italy, GI in Germany, the British Computer Society in the UK, IEEE and ACM in the US, etc.). Its activity is run by technical committees and by special interest working groups.

One of the first books that Corrado gave me to read (which contained one of his landmark papers on the language CUCH) was *Formal Language Description Languages for Computer Programming* and consisted of the proceedings of a working conference held in Vienna by IFIP TC2 in September 1964. In this book several papers were devoted to the presentation and discussion of proposals for meta-languages able to describe programming languages from the syntactic point of view (but some papers, notably the paper *Towards a Formal Semantics* by Christopher Strachey, also attempted to show how a metalanguage could capture semantic features of a programming language[19]).

All aspects of this blooming field of computer science were present at the NATO school in Pisa. As I already said the lecturers were some of the most prominent computer scientists in the world.

Seymour Ginsburg (born in 1927 and died in 2004), a mathematician, was one of the founding fathers of formal language theory in the US (see Fig. 1.6). In 1965 he was working at SDC (one of the world's first software companies). His research group included Sheila Greibach, Michael Harrison, Jeff Ullman, all names very well known for their contributions in the field. At the time of the NATO school in Pisa the major contributions of Ginsburg were related to the theory of context-free languages, a domain in which, in particular with Sheila Greibach, he established groundbreaking results. A couple of years later Ginsburg and Greibach founded the theory of Abstract Families of Languages (AFLs), a mathematical theory developed by abstraction, starting from the properties of regular languages and of context-free languages, that contributed to the dissemination of algebraic methods in computer science. After the SDC years Ginsburg moved to the University of Southern California where he remained until retirement, working in various other domains of

Fig. 1.6 Seymour Ginsburg

---

[19] The paper *Towards a Formal Semantics* is indeed considered to be the beginning of the development of the approach to semantics known as mathematical semantics or denotational semantics.

Fig. 1.7 Peter Naur

theoretical computer science, including database theory. His lectures in Pisa had the properties of context-free languages as the central topic. What is now textbook material, such as the notion of ambiguity, the pumping lemma (that allows us to characterize context-free languages), the Greibach normal form (that is the basis for the definition of various compiling techniques), were at that time recent results that Seymour Ginsburg presented, usually with lots of details and technicalities, while an expression of real satisfaction appeared on his large, chubby face.

Adriaan van Wijngaarden and Peter Naur had both been involved in the definition of Algol 60 and both had taken part in the famous Paris meeting. In particular Peter Naur (Fig. 1.7) had the role of editor of the influential *Report on the Algorithmic Language ALGOL 60*. His role in the definition of Algol 60 was one of the major motivations that led the ACM to grant him the Turing Award in 2005. Born in 1928, originally he was an astronomer but he turned to computer science when he worked at Regnecentralen, the first Danish computer company, from 1959 to 1969. Subsequently he was a professor of computer science at the University of Copenhagen until 1998. His work was mainly related to the syntactical definition of programming languages. For its relevance, the BNF (Backus Normal Form), the metalanguage with which we currently define the syntax of programming languages (that took its name from John Backus, the American computer scientist who invented it in 1959, and that was used for the first time in the above-mentioned report), is often called Backus-Naur Form.

In his course on the systematic design of effective compilers Peter Naur presented us with the design of the Algol 60 compiler for the Gier machine[20],

---

[20] The Gier machine was manufactured by Regnecentralen in Copenhagen and was equipped with a main storage with 1024 words of 42 bits plus a magnetic drum consisting of about 12,000 words.

developed in Copenhagen in 1962. His paper *The Design of the Gier Algol Compiler*, provides a very interesting overview of the difficulties that had to be faced in such early stages of compiler construction. It is interesting from this point of view to see what Naur said about the contemporary effort of Edsger Dijkstra at the Mathematisch Centrum in Amsterdam and his own difficulty in following the same direction: "The Dutch group impressed us greatly by their very general approach. However although they were prepared to put their solution of the problem of recursive procedures at our disposal we decided to stick to the more modest approach which we had already developed to some extent. The reasons for this reluctance were practical. First of all we felt the size of the problem to be already rather frightening ...".

Also Adriaan van Wijngaarden[21] (born in 1916 and died in 1987) originally was not a computer scientist. He was a mechanical engineer but already in 1947 he became the head of the Computing Department of the newly founded Mathematisch Centrum, the computing research center in Amsterdam that he was directing in 1965 at the time of the NATO school and that he directed until 1981. His research work was devoted for several years to the definition of formal metalanguages for the specification of programming languages, in particular the so-called two-level grammars that he applied to the definition of the language Algol 68. His lectures are still impressed in my mind for the clarity and crispness of his words accompanied by his sharp voice.

Finally, among the speakers there was another outstanding scientist: Peter Landin (Fig. 1.8). Peter Landin (born in 1930 and died in 2009) was a British computer scientist who had the intuition to bring into programming languages concepts and constructs inspired by λ-calculus. From 1960 to 1964 he had been an assistant of Christopher Strachey. His work had a great influence both on the design of functional programming languages and on the subsequent developments in the study of the semantics of programs. At the school in Pisa he gave an introduction to λ-calculus and presented a programming language in which the principles of λ-calculus were embedded, together with an abstract machine (the SECD machine) for its interpretation.

In all the lectures what I found fascinating was the effort of syntactic formalization that was leading the search for metalanguages, concepts that were totally new to me. I was assimilating, day by day (the school went on for three weeks), the new mathematical ideas that were disclosed to me, a mathematical world so different from the mathematics I had learned in my calculus classes. Beside the lectures, the school (as always happens) was also offering me a great occasion to meet some of the young scientists who were starting to appear in the European computer science scene. Among them were Jaco de Bakker, who later would become one of the major experts in

---

[21] For the fundamental role that this scientist had in developing computer science in The Netherlands and in the history of CWI, in 2006 CWI created the van Wijgaarden Award that is assigned every 5 years to distinguished computer scientists and mathematicians.

Fig. 1.8 Peter Landin

program semantics[22], Antonio Grasselli, a chemical engineer graduated at Milan Polytechnic, later one of the founding fathers of the computer science department in Pisa, and Fabrizio Luccio, an electrical engineer, at that time assistant professor at Milan Polytechnic. After a period of research activity in the States Luccio was hired in Pisa and, as we will see, our lives crossed over again.

On the weekend between the second and third week of the school I was asked by Corrado Böhm to attend a meeting in Padua. The meeting was held by the Committee of Analog Computers of the Italian Association for Automatic Computing (AICA) in conjunction with a conference on Analog Sequential Computers. At that time the committee was chaired by Antonio Ruberti, professor of Electrical Engineering at Rome University[23].

The meeting was essentially devoted to discuss recent developments regarding hybrid computers and Corrado wanted to catch this opportunity to announce, at least informally, his aim (the task I was assigned) to define a programming language for hybrid computers. The topic raised some interest among the members in the committee, although it was not clear whether the questions posed to Corrado were witnessing real interest or simple courtesy toward him. In any case this was the first time in which I heard the subject of my work being presented to a scientific audience. It was also the first time

---

[22] Jacobus Willem (Jaco) de Bakker (1939–2012) took his PhD under the supervision of Adriaan van Wijngaarden and from 1964 he was one of the research leaders of the Mathematisch Centrum, a pioneer in the field of program correctness and verification and of program semantics.

[23] Antonio Ruberti (1927–2000) was one of the leaders, actually the most important leader, of the Italian school of control and system theory in the last century. As Italian Minister of University and Research and as EU Commissioner for Research he made a deep political and cultural impact in Italy and Europe. Also his name will occur several times in these notes.

I met Antonio Ruberti, who ten years later would play such a big role in my life.

After a short excursion in Venice I went back to Pisa, back to programming languages and metalanguages.

In the last week at the school the lectures given by Caracciolo came. Those lectures really had a strong impact on the work I was carrying out for my thesis. In his lectures the 'Prince' was giving the formal definition (in BNF) of his programming language for machine tools: all of a sudden I realized the affinity between the programming language discussed in Caracciolo's lectures and my programming language for hybrid computers. In both cases the programming language should contemplate a digital control over the application of continuous functions and continuous operators. An immediate consequence of this affinity was that the metalanguage used by Caracciolo could have also been applied to provide the formal definition of a programming language for hybrid computers. The track for my work was drawn.

After the school I went back to Rome and started to write the body of the thesis: the description of my CUCH-based language for hybrid computers. After an introduction regarding the pros and cons of digital computers versus analog computers and the technical characteristics of hybrid computers, the thesis contained a note on the metalanguage (a slightly modified version of BNF), a description of the 'object language' LISSA (the language I had defined in order to describe the main structural components and the physical layout of a hybrid machine) and an outline of the 'translator' (the software, based on the reduction mechanism of λ-calculus expressions, that was supposed to transform the high-level formulation of a mathematical problem into a description of the physical layout of the hybrid machine required for its solution).

In January the write-up was finished and in February 1966 I graduated with 106/110. The discussion of the thesis was a bit surrealistic; the 'controrelatore' (discussant) was Romano Bizzarri, professor of particle physics, and although it was clear that he considered the topic of my thesis rather unusual for a degree in physics he was curious enough to raise appropriate questions to which I gave apparently valid answers.

I was a Doctor in Physics.

Chapter 2
Lots of Insipid Stupid Parentheses

In 1966 the competition for obtaining a scholarship of CNR (the Italian National Research Council) was not as tight as it is now. Five scholarships were announced by INAC and only three candidates submitted their curriculum and their thesis. Encouraged by Corrado Böhm, I was one of them. My choice to try a scientific career was already clear to me since at least two years before. In 1964 I had attended a summer school organized by IBM at Rivoltella del Garda. From that school I still have good memories of the introductory lectures regarding computers, programming languages and operations research and, above all, I have wonderful memories of Sirmione and of Lake Garda! What I did not like was the fact that for the IBM engineers who were lecturing at the school the main interest was only in computer applications. What I did not like for sure was their stiff look and the motto 'Riflettete!' (the Italian translation of the well-known IBM motto 'Think!') written everywhere, even in the bedrooms. So at the end of the school my choice was clear. I would never take a job in industry. When I talked with IBM officials at the end of the school, about being hired by IBM after graduation, I suggested instead my interest in working in a research institution (university or CNR).

At the beginning of April 1966 I started to work at INAC with my scholarship and the first task that was assigned to me was to study the manual of a new computer that had been installed at the institute: the Olivetti-GE Elea 9104. This machine had been designed at Olivetti as a result of an agreement established in 1961 between the leader of the electronics projects of Olivetti, Mario Tchou, and INAC's former director, Mauro Picone. Two INAC's researchers took part in the project: Roberto Vacca (who left CNR a short time after) and Paolo Ercoli.

Paolo Ercoli was one of the founding fathers of Italian computer science. Born in 1926, after graduating in Electrical Engineering and after working for some time in electrical and telephone companies, he was hired by CNR in 1955 and in July 1956 became Director of the computing center of INAC. Since 1956, in collaboration with Roberto Vacca, he gave an introductory course on electronic computers at the Faculty of Engineering, a course that

G. Ausiello, *The Making of a New Science*,
https://doi.org/10.1007/978-3-319-62680-2_2

in 1960 became the first official computer course in Rome University; Paolo
Ercoli was officially in charge of this course until 1974 when he became full
professor. His role in Rome University (as also in my academic life) was
fundamental until his death in 1996.

Fig. 2.1 The CINAC computer. In front the emulator of FINAC

The machine (to be installed at INAC and dubbed CINAC, Calcolatore
INAC) was based on transistors technology and was conceived with advanced
computer hardware design principles (interrupt, multitasking, dynamic storage allocation and a stack fast memory) that made it particularly suited for
executing Algol programs (see Fig. 2.1).

But Elea 9104 was not born under a lucky star. Mario Tchou had died in
a car accident in November 1961 (at the age of 37), and a few years later,
in 1964, the Electronics Division of Olivetti had been sold to a company
owned by General Electric (75%) and by Olivetti (25%)[1]. As a consequence
the ambitious Elea 9104 project was never completed.

The main problem of the new machine was at the software level. Almost no
piece of system software (except the loader) had been developed by Olivetti,
not to speak about the compiler for the high-level programming language
(a variant of Algol called Palgo). Therefore the Director of INAC (Aldo
Ghizzetti, a mathematician) created a group with the task to implement,
first of all, the operating system and a symbolic assembly language. The
group was led by Paolo Ercoli, Vice Director of INAC, and consisted of various young researchers: Giuseppe Iazeolla, Puccio Jacopini, Marisa Venturini
Zilli and a few others. I was asked to collaborate with this group.

The machine was programmed in 'octal' machine language. It is hard to
imagine now: any three bits corresponded to an integer between 0 and 7.

---

[1] This share was doomed to become 100% General Electric four years later. In 1960 Adriano
Olivetti, the visionary leader of the company, had died, and when in 1964 Olivetti, suffering
a financial crisis, opened its capital to banks and Italian manufacturers, the new partners
decided to give up the competition in the computer market, blindly obeying a clear dictat
of the international labor division established at the end of WWII.

A byte consisted of two octal digits and a word consisted of four (or eight) bytes (24 or 48 bits). The whole memory of CINAC contained 98,304 24 bits words. Machine instructions consisted of four octal digits (e.g. 7422 was the code for a data transfer instruction) plus four octal digits to address a memory location in a page of 4096 words. Writing a program was a nightmare. A program was written with a pencil (and an eraser!) on a sheet of paper and then typed on a punched tape that was finally loaded into the memory by means of an optical device.

The machine was also equipped with a simulator of FINAC, the old Ferranti machine. This was meant to allow the use of the FINAC legacy software and also to allow CNR programmers to maintain their own programming environment: the console and the old I/O devices. Even the sound that the old machine used to emit while processing data was reproduced in order to preserve also the same physical environment for the programmers. So, while the team of developers was working to implement the missing system software, INAC mathematicians could keep their normal numerical data processing activity.

I was sitting in a room with two friendly colleagues: Mario Rosati and Paolo Bulgarelli, two real mathematicians who were doing research in analysis, in particular on the solution of free boundary problems, in collaboration with other CNR researchers. My advisor was of course Corrado Böhm with whom I had the privilege to learn, day by day, new problems, new concepts, new aspects of theoretical computer science. With Corrado I wrote my first scientific paper, an INAC report with the description of the LISSA programming language for hybrid computers that I had designed in my master's thesis. The paper was submitted to a conference (the IX Convegno sull'Automazione e Strumentazione) and at the end of November I went to Milan where I presented my CUCH-based programming language for hybrid analog computers. That was the first time I was presenting my work at a conference. I still remember: I was explaining my work, in the dark of the room, writing on an overhead projector, in front of fifty some people, mostly hardcore engineers who had never before heard about λ-calculus. In a sense I was feeling as if I belonged to a new generation. Frightening but exciting!

As I said my task at INAC was to take part in the implementation of the operating system for the CINAC machine. In particular I was asked to write the drivers (the sofware that 'drives' the input and output devices). Besides I started to join Puccio Jacopini and Marisa Venturini Zilli as a teaching assistant in Corrado's course of 'Programming techniques' and I used to follow all Corrado's lectures and to correct the homework of his students.

In June 1966 I got married to my first wife, Patrizia, and I went to the Dalmatian Coast for the honeymoon. At that time no electronic mail, no cellular phones. For two weeks the work contacts were switched off. λ-calculus faded away.

As soon as I went back to Rome Corrado Böhm suggested I attend an important event that was taking place in July at the Hotel Parco dei Principi, in Rome: the *Journées internationales d'étude sur la théorie des graphes et ses applications*. The fact that the title was in French should not be surprising. The conference was organized by the International Computation Centre that, from 1965 to 1967, was directed by the French mathematician Claude Berge[2], who in a sense can be considered the father of Graph Theory.

The International Computation Centre had been established, first in a provisional way, by a bilateral agreement concluded in September 1957 between UNESCO and the Italian Institute of Higher Mathematics (Istituto Nazionale di Alta Matematica), and continued its activity until 1970 when it was moved to Geneva. The main function of the centre was "to ensure mutual assistance and international collaboration between existing bodies dealing with computation and information processing, in particular as regards scientific and technological studies and to help, on request, the countries which do not possess their own computation equipment". In a document of 1961 regarding the prospects of the International Computation Centre it is said that the main activity would have concerned the solution of large numerical problems and the treatment of data deriving from meteorology and census. The arrival of Claude Berge as Director had the effect to move the interest of ICC toward discrete mathematics, computer science, and operations research, and the idea to organize a conference on graph theory was clearly due to this shift of interest.

The conference was one of the first events devoted to this new branch of discrete mathematics, certainly the first organized in Italy. How recent was the interest in this discipline is revealed by the introductory words of Claude Berge (Fig. 2.2): "Au nom du Centre International de Calcul, je vous remercie d'être venus si nombreux à Rome pour un sujet qui, ils y a dix ans seulement, n'aurait attiré qu'une dizaine de personnes". Taking part in the conference I learned for the first time about graphs and their applica-

Fig. 2.2 Claude Berge

---

[2] Claude Berge was born in 1926 and died in 2002. His book *Théorie des graphes et ses applications*, written in 1958, has been for several decades the major standard reference book in the field.

tions (again a type of mathematics completely absent from our university curricula) and the discovery was exciting. I was fascinated by those strange mathematical objects whose strong modeling power had been explained by Berge: "L'étude systématique des graphes a permis de relier entre eux des résultats et des théories sans liens apparents. Le langage des graphes s'est révélé le plus adéquat pour un grand nombre de problèmes combinatoires, et cela parce qu'on commence à apercevoir une théorie cohérente. Les problèmes distincts se détachent, les problèmes analogues se rejoignent, et les lacunes se comblent."

Then, in September another NATO Summer School came; the *Advanced Study Institute in Programming Languages 1966* was organized by François Genuys in Villard de Lans (Grenoble). I attended the school and again this gave me the chance to meet some of the main figures of the computer science scene in that moment. This time the main lecturers of the school were Ole-Johan Dahl, Edsger Dijkstra, Tony Hoare, Robert Floyd; how prestigious was the school is proven by the fact that all four of them would receive the Turing Award in subsequent years! The other speakers were Calvin Elgot, researcher at the IBM Research Center in Yorktown Heights, who lectured about an abstract formalization of flow diagrams, and Louis Bolliet, professor in Grenoble, who spoke about principles of compiler construction. A few lines to describe the lecturers and the content of their lectures will explain why, as students, we had the feeling that sitting in the classroom we were assisting in some fundamental steps in the foundation of the new discipline.

Ole-Johan Dahl (Fig. 2.3) was at that time a researcher at the Norwegian Computing Center. Together with another Norwegian scientist, Kristen Nygaard, he had been engaged from 1961 to 1965 in a big project: the design of a simulation language, called SIMULA I. At the school he lectured about simulation languages for discrete event systems (making a comparison between SIMULA and other simulation languages such as GPSS) and described the developments that were leading to the design of an advanced version of SIMULA, later called SIMULA 67. SIMULA I and SIMULA 67 are generally considered the first examples of one of the most important families of programming languages, the 'object-oriented' languages. During his lectures, for

Fig. 2.3 Ole-Johan Dahl (left) and Kristen Nygaard at the time of Simula's development

Fig. 2.4 Edsger Dijkstra

the first time, I heard the terms 'object', 'class of objects', 'instance', concepts that twenty-five years later became the basis of modern programming languages such as C++, C# and Java[3].

Edsger Dijkstra (Fig. 2.4) lectured about co-operating sequential processes. Since the early days of computing the problem of concurrency control has been a major research issue in many contexts: operating systems design, multiprocessing and real-time processing (such as in process control, traffic control, banking applications). Again in Dijkstra's lectures we were exposed to the formalization of concepts that later would become cornerstones of computer science education: the concept of 'critical section' to indicate activities that should be carried on by different processes in 'mutual exclusion' or the concept of 'semaphore' (the signaling tool that Dijkstra introduced with its primitives P and V borrowed from the Dutch railroad terminology) and finally the concept of 'deadly embrace' (deadlock).

The paradigms of the dining philosophers (referring to the competition among processes for the use of shared resources) and of the producer and consumer (referring to the cooperation among processes in the production and consumption of resources) have since become classical in operating system courses. In the 1970s and 1980s the field of concurrency really exploded and various other formalisms for defining properties of concurrent systems (first of all Hoare's communicating sequential processes) were introduced and opened the way to process calculi, process algebrae and to the constructs for concurrency control that were later included in programming languages.

---

[3] Dahl died in 2002, at the age of 71, one year after having received the Turing Award (jointly with Nygaard) for having invented the language SIMULA and for having introduced the concepts of object-oriented programming.

The title of the lectures given by Tony Hoare[4] (Fig. 2.5) (at that time working at the British computer company Elliott Brothers) was *Record Handling in Programming Languages*. In a sense his lectures were correlated with Dahl's but they were focused on the logical and physical concept of record and provided a proposal to extend Algol 60 with structured data. It turned out to be a good preparation for the students attending the school toward the programming languages that would be developed a few years later (Algol 68, PL/I, Pascal).

Fig. 2.5 Charles Anthony (Tony) R. Hoare

Finally, Bob Floyd[5] (Fig. 2.6). After working a few years in hi-tech companies Bob Floyd had become associate professor at Carnegie Mellon. At the NATO School he lectured about program verification, the subject on which he was working in that period. His approach was based on establishing conditions that should hold after the execution of a statement on the proviso

[4] Sir Charles Antony (Tony) R. Hoare (born in 1934) was assigned the Turing Award in 1980 "for his fundamental contributions to the definition and design of programming languages". Hoare is also famous for having invented Quicksort, a popular sorting algorithm with average running time $O(n \log n)$ (that he conceived in 1960 while he was working on automatic translation at Moscow University), and for having developed a logic system for proving correctness of structured programs, known as Hoare's logic.

[5] Robert W. (Bob) Floyd (1936–2001) was a professor of computer science at Stanford from 1973 to his death. In 1978 he won the Turing Award. He is also famous for having invented a popular algorithm for computing the shortest paths between all pairs of nodes in a graph, following the same dynamic programming approach used by Stephen Warshall for computing the transitive closure of a directed graph. The citation of the award reveals how wide and deep was his contribution to computer science: ".. for helping to found the following important subfields of computer science: the theory of parsing, the semantics of programming languages, automatic program verification, automatic program synthesis, and analysis of algorithms".

Fig. 2.6 Robert W. (Bob) Floyd

that a suitable condition was holding before the execution of the statement, in the direction that later would become known as 'Hoare's logic'.

During the school we had the chance to make some wonderful excursions in the mountains around Grenoble and I also met good friends. An Iranian student that I had already met the year before in Pisa, Farhad Mavaddat (later professor at the University of Waterloo) was my companion on pleasant hikes, and another Italian student Stefano Crespi-Reghizzi has since then become a dear friend. I also remember a French student, Alain Colmerauer, who would later become famous for inventing the programming language PROLOG, and a Danish student, Dines Bjørner, who would become the great boss of 'formal methods' (for program design and program verification) in the 1980s and 1990s.

When I went back to Rome I realized that the design of system software for CINAC was proceeding rather slowly. Every week the team used to meet together with INAC's Director, Aldo Ghizzetti, but the meetings were revealing different views in the group. Corrado Böhm, taking part in the meetings, used to repeat 'Erigenda meridiana'. The words (inspired by those that Cato used to pronounce in front of the Roman Senate: 'Delenda Carthago', by which he wanted to remind Romans of the urgency of stopping the maritime power of the strong rival of Rome) meant that it was urgent to create time-division facilities in order to allow multiprogramming on CINAC. In principle the machine was equipped for such developments, since the machine language included instructions for dynamically relocating software, but the way to go to realize a suitable operating system was still very long.

My task was supposed to involve the implementation of the drivers for the machine but Corrado had a surprise for me. He wanted to get me involved in a visionary project: the construction of an alternative software environment on the CINAC machine, a LISP-based environment.

LISP is a functional programming language that derives its name from the words LISt Processing; linked lists are in fact the major data structure together with integers and the so-called atoms (identifiers with attached properties). Through its various dialects (more than thirty!) it is the second-oldest programming language still in use today due to the great success it had in the AI community (but not only). Ideas and concepts introduced in the language are still widely used in contemporary languages (both functional, such as, for example, Scheme, Clojure or Haskell, and non-functional, such as Python, Ruby or Scala) and computational frameworks, for example the well-known Map Reduce framework for parallel programming introduced by Google to deal with big data management and computing derives from the LISP meta-functions map and reduce (where "meta" means that they receive functions among their parameters).

LISP had been invented by John McCarthy (Fig. 2.7) at MIT in the late 1950s. In his words: "My desire for an algebraic list processing language for artificial intelligence work on the IBM 704 computer arose in the summer of 1956 during the Dartmouth Summer Research Project on Artificial Intelligence which was the first organized study of AI". In the Fall of 1958 McCarthy had become assistant professor of Communication Sciences in the Electrical Engineering Department of MIT, where he and Marvin Minsky had started the Artificial Intelligence Project[6]. LISP was not the first programming language for non-numerical computing but it was the first such language based on a sound theoretical basis, borrowing from Church's λ-calculus and from Kleene's recursive functions. In 1960 John McCarthy wrote a paper, *Recursive Functions of Symbolic Expressions and their Computation by Ma-*

Fig. 2.7 John McCarthy

---

[6] See *History of LISP*, by John McCarthy, 1979 (www-formal.stanford.edu/jmc/history/lisp). Both Marvin Minsky (1927–2016) and John McCarthy (1927–2011) received the Turing Award, respectively in 1969 and in 1971 for their role in creating and advancing the field of Artificial Intelligence. In particular Marvin Minsky is also considered the father of machine learning by means of neural networks. In the book *Perceptrons*, that he wrote in 1969 together with Seymour Papert, Minsky studies strong aspects and weak points of a particular kind of neural network, the Perceptron, invented by Frank Rosenblatt in the late 1950s. Minsky also became famous for inventing (together with Papert) the programming language LOGO, particularly suited for teaching algorithmic concepts to children.

chine, in which LISP is presented as a "formalism for doing recursive function theory" and is proved to be Turing-complete (i.e. to be as powerful as Turing machines in terms of computing power, in the sense that any algorithm implemented by a Turing machine can be implemented also in LISP).

After the first version (LISP 1), in 1962 version 1.5 was developed for the IBM computer 7090, again by John McCarthy at MIT. LISP 1.5 was soon widely distributed. Immediately the success of the language was reflected by a variety of applications in all fields of non-numerical processing. One year later a new interesting version was implemented for the PDP-1 machine (one of the first so-called minicomputers) by Peter L. Deutsch (at that time only 16 years old!) and Edmund C. Berkeley. With respect to the 7090 version, the PDP-1 implementation was more flexible and, most important, more interactive: instead of providing the whole program as a deck of cards before the computation as in the IBM version, expressions could be typed by the user on a keyboard while the output was typed automatically by the typewriter in what became later the standard way of operating in a LISP environment.

The interest of ARPA, the Advanced Research Projects Agency of the Department of Defense, shows how strong were the expectations of advanced research institutions for the new programming language and for the applications of non-numerical computing. The agency supported a joint initiative with Project MAC and with the private company Information International Inc. for the development of LISP and for its dissemination through the publication of the MIT Press volume *The Programming Language LISP: Its Operation and Applications* edited by Edmund Berkeley and Daniel Bobrow.

The MIT book and later the green volume written by Clark Weissman at the California software company SDC, *LISP 1.5 Primer*, were my first readings regarding LISP when, in Fall 1966, the Mexican scholar Hugo Moncayo Lopez arrived at INAC, invited by Corrado Böhm. Hugo had taken part in the development of a LISP interpreter designed by T.A. Brody in 1965 at the National Computing Center in Mexico City for the machine IBM 1620 (called Lispito) and Corrado asked him to write a LISP interpreter for CINAC. LISP made its entrance at INAC, better, LISP made its entrance in Italy.

The first Italian LISP interpreter (dubbed ROMALISP) was written by Hugo Moncayo Lopez in three months with my help, under the direction of Corrado. Considering that the interpreter was written in machine language (the above-mentioned octal language), that the debugging required us to 'cut and paste' (literally!) the punched tape, and that the programming environment was a rudimentary system designed on purpose to support the LISP interpreter (allowing us to load programs, to dump the whole CINAC memory and to perform a few other operations like accessing magnetic tapes, putting a program in pause and resuming the computation etc.), the development of ROMALISP was a miracle. In February 1967 we had a high-level programming language available on the CINAC machine.

Soon the LISP interpreter became our main research interest. Due to its mathematical notation and despite the problem of dealing with expressions with deeply nested parentheses[7], writing LISP programs is amusing, much more than writing programs in any other programming language. Among the first toy applications I remember programs for processing DNA sequences and for analyzing logical formulae. Some applications were also conceived in collaboration with Domenico Parisi, CNR researcher at the Institute of Psychology. Unfortunately processing non-numerical information was space consuming and our applications could not be so large and powerful. As I said, CINAC used to communicate with the outside world not only through I/O devices but also using sound. The execution of LISP programs played a characteristic 'music' and in particular the 'garbage collector' (the LISP system software component that has the task to discard unused linked lists and to retrieve free storage space and make it available again) produced a typical sound: when a large, memory consuming program was executed the sound of the garbage collector used to become more and more frequent meaning that the computer memory was becoming exhausted.

In any case after the toy applications we started to address more important issues. The first one was the research subject of Alfonso Miola, a student in mathematics who was going to start his master's thesis under the supervision of Corrado (and who later became professor at the University of Roma Tre): the creation of an algebraic manipulation system.

Algebraic manipulation systems were conceived to allow us to process algebraic expressions at the symbolic level. While in a numeric computation system the instruction $x \leftarrow (a + b)(a - b)$ is supposed to mean: "retrieve the numerical values of a and b and compute the numerical value of x", in a system capable of performing symbolic computations the result of the instruction would be to assign as value of x the expression $a^2 - b^2$. Similarly, this type of system allows us to perform the symbolic derivative of an expression (e.g. $d(x^2 + 3)/dx = 2x$). Although the first attempts to create this type of processing system started in 1953, only in the 1960s, when LISP became available, was it possible to realize powerful and practical algebraic manipulation systems. This type of application, actually, was for a while one of the major directions of artificial intelligence research.

In 1964 Carl Engleman at MITRE Corp.[8], in Massachussets, developed MATHLAB, still today a widely used integrated system for doing symbolic and numeric computing, in LISP. In 1966, REDUCE, another computer algebra system, was developed at Stanford, by Anthony Hearn. Hearn's original motivation was to perform the symbolic computation of Feynman diagrams, and meeting with John McCarthy, just arrived in Stanford, he understood

---

[7] Several jokes have been made around the tedious use of parentheses in LISP. From this derives one of the explanations of the acronym LISP: Lots of Insipid Stupid Parentheses.

[8] In the early 1960s several hi-tech companies, such as MITRE Corporation and Computer Associates, were located in the Boston area, close to MIT, a source of advanced ideas and of skilled researchers. One of them, BBN (Bolt, Beranek and Newman), very active in AI research and in the ARPA net development, was called "the third university of Cambridge" (after MIT and Harvard).

that LISP was the right programming language to achieve his purpose. Finally in 1968 Macsyma (Project MAC's SYmbolic MAnipulator) was developed, again in LISP, at MIT as part of Project MAC by Carl Engleman and Joel Moses. Subsequently dozens of researchers cooperated on the project, developing a huge library of functions (ranging from formal derivation and integration to tensor calculus and polynomial factorization). Many functions and ideas of Macsyma were later incorporated in other systems such as Mathematica and Maple, developed in the 1980s and probably now the most commonly used computer algebra systems.

The algebraic manipulation system developed by Alfonso Miola at INAC (called SINAC) was the first real application of ROMALISP and the first such system implemented in Italy. Several mathematics students were engaged under our guidance to develop new functions for the system or to investigate theoretical aspects of symbolic computing (such as the undecidability of establishing whether certain elementary functions are identically zero).

In May 1967 I got a permanent position as researcher at INAC with the official title of 'Aspirante ricercatore' (a title that familiarly might sound like: 'wannabe researcher'). My principal task was to work on the LISP system and its applications. About at the same time, in order to pursue his ambitious program, Corrado Böhm had invited another foreign student (as we have seen, a particularly gifted student!), Peter Deutsch (Fig. 2.8), to come to Rome. Peter was born in Boston in 1946. His father, Martin, was professor of physics at MIT and was very well known for having discovered positronium, a system consisting of an electron and its anti-particle, a positron. In 1967 Peter was a student in Berkeley and was exactly as you would imagine a student in Berkeley in the late 1960s: long hair, short pants, walking barefoot. I went to the air terminal to pick him up and I was anxious to talk to him about the work he was supposed to carry on at INAC, but he was not alone. He was with a girlfriend (again you can imagine her as coming out of a Woodstock concert photo) and they just wanted to be taken to the hotel.

As I mentioned before, Peter had developed a LISP 1.5 implementation for PDP-1 when he was still a teenager. The task that Corrado assigned him was to make a LISP implementation of a compiler for a FORTRAN-

Fig. 2.8 Peter Deutsch

like programming language that might be used by INAC mathematicians to carry on their usual numerical computing work: in this way, exploiting the power of LISP, INAC would make a leap into the future, coming out from the anguish of the never-ending construction of system software for the CINAC. Peter Deutsch had a furious approach to programming[9]. I remember him running barefoot in the corridors of the institute. Two or three times per day he used to come to my office to ask for some upgrade or add-on to the ROMALISP interpreter or to the operating system. After two months the symbolic assembler and the compiler of an imperative language reminiscent of PL/I were ready. Their power was limited by the fact that they were executed on top of the LISP interpreter so they were rather slow and, more important, sometimes they halted because of an overflow of the interpreter stack.

When Peter Deutsch left he gave the draft of a revised version of ROMA-LISP to me and Alfonso, and suggested we rewrite the interpreter according to his recommendations, in order to make it more efficient. An important improvement was to introduce the so-called 'program feature' that allows us to perform iterative computations still in a recursive environment. Also, the garbage collector should be redesigned by making use of a pointer reversing technique for the implementation of the depth-first search algorithm, instead of accumulating pointers in the push down store: in such a way we would eliminate the frequent memory overflows that were limiting the power of our interpreter. Alfonso and I accepted with enthusiasm the task and in a few months the new interpreter, ROMALISP 2, was ready.

In Fall 1967 another event had an important role in my education: Corrado Böhm left for a three-months trip to the US and I was asked to replace him as lecturer during his absence. I had to start the course in November and to give the first ten or twelve lectures until the beginning of the Christmas vacation. My role at that moment was teaching assistant: in the bureaucratic language of the Italian university in those years, 'voluntary assistant' since I volunteered to help to teach and to assign and correct homework.

It was the first time I was giving real lectures and although I had the notes of the course given by Corrado the year before I was really worried about such a hard task. I plunged into the study of more advanced material regarding computability and formal languages. In particular I went with enthusiasm through one of the 'holy books' given to me by Corrado: Martin Davis' *Computability & Unsolvability*, a wonderful book whose content spans from mathematical logic to a systematic account of various approaches to computability, and that provides the reader with a very detailed treatment of Turing machines. I also had copies of some of the first papers written by Chomsky, in particular the papers *Introduction to the Formal Analysis of Natural Languages* and *Formal Properties of Grammars*, two chapters of the

---

[9] The contributions that Peter Deutsch gave to software systems in the subsequent years are very important and range from the creation of Ghostscript (a Postscript and PDF interpreter) to a noteworthy implementation of the programming language Smalltalk.

Handbook of Mathematical Psychology, published in 1963, where the properties of Chomsky grammars and of the corresponding families of languages were explained in a clear and systematic way (no book on the subject had yet appeared in those days).

The paradigm on which I based my lectures, deriving from what I learned from Corrado, was the similarity between logical theories, rewriting systems and abstract machines. Corrado used to consider the definition of rewriting systems by the Norwegian mathematician Axel Thue[10] in 1914 as the real beginning of the history of computer science and the correspondence that could be established between the concept of proof in a logical system, of derivation in a rewriting system, and of computation by a Turing machine was an elegant approach to introduce decidability, computability and formal languages in a unified framework.

The course was attended by few students but since the first lecture I realized that I should be very well prepared: some of the students were ready to catch any hesitation or any lack of precision in what I was saying. Among them two were particularly reactive: Marco Marfurt and Bruno Simeone. Later they would become professor of numerical analysis and professor of operations research respectively[11]. I enjoyed teaching, especially I enjoyed preparing the lectures and conveying what I believed to be organized and convincing arguments to the students. A pleasure that did not leave me throughout the rest of my life.

Corrado came back in December 1967. While the development of ROMA-LISP and its applications were going on, the research work on λ-calculus at INAC was very active. Under the leadership of Corrado Böhm and Wolf Gross the research group included Puccio Jacopini, Marisa Venturini Zilli, Alfonso Miola and myself. In that period Corrado was traversing a state of exceptional scientific productivity. In 1968 he produced one of his major achievements: he proved that within λ-calculus it is possible to 'separate' two syntactically distinct normal forms, that is, in a suitable context they can be reduced to two preassigned terms, an important result later known in the literature as 'Böhm's Theorem'. Another subject to which Corrado devoted his attention was the algorithmic aspect of λ-calculus and the evaluation mechanisms of λ-expressions. In this context Corrado proposed to me another major project: the design and implementation of an abstract machine for playing with λ-calculus on the LISP interpreter.

---

[10] Axel Thue (1863–1922) was a Norwegian mathematician who made fundamental contributions to algebra and combinatorics. In particular he defined the word problem for semigroups (given a semigroup and given two words u and v, can u be transformed to v by using the rules of the semigroup?) that later was proved to be undecidable. Thue's ideas and results can be considered the first contributions to the fields known today as combinatorics of words and coding theory.

[11] Bruno Simeone was indeed one of the most important figures of Italian operations research, specialized in quadratic optimization and in electoral systems. Bruno died in 2010 at the age of 65.

The problem I had to approach had been already addressed by various authors. Peter Landin, in particular, had defined his SECD machine to evaluate λ-calculus expressions interpreted as numeric functions, and, more generally, to interpret a functional programming language. The acronym derived from the names of the internal registers: Stack (pointing to a list of intermediate evaluation results), Environment (pointing to a list of < *name, value* > pairs), Control (a list of expressions waiting for evaluation), Dump (pointing to the content of the registers in the preceding state). Other authors had taken a completely different way, based on a purely syntactic approach, in which the efficiency of the evaluation mechanism was totally ignored. All such research efforts were, on one side, early attempts to formalize the run-time behavior of programming languages and, at the same time, were the first studies in which λ-calculus revealed itself as a powerful tool for modeling the semantics of programming languages, a target that was reached thanks to the work of Christopher Strachey, Dana Scott and Gordon Plotkin.

The version I implemented on the ROMALISP 2 interpreter was inspired by the LISP evaluation mechanism but at the same time was essentially syntactic and was not tailored to any prespecified interpretation of CUCH-expressions. The system was described in a paper that we submitted to the AICA Conference (the annual conference of the recently founded Italian Association for Automatic Computing[12]) that was going to be held in Naples in October. The paper was accepted. The Session Chair, who at the end asked a few questions, was Massimiliano Lunelli, a mathematician from Milan Polytechnic, one of the few Italian mathematicians who was sincerely interested in the development of computer science.

But 1968 was not an ordinary year. In Paris, the students movement had disrupted the French academic system and upset French society with revolutionary words: "l'imagination au pouvoir", "ce n'est qu'un debut continuons le combat". In Rome, on March 1st a violent clash between students and police took place in front of the Faculty of Architecture, at Valle Giulia, and triggered occupations and sit-ins in various universities around Italy.

The Italian National Research Council was not exempt from such political upheaval. CNR researchers (a couple of thousand at that moment) wanted to join the demand of university students for greater investments in education and research and were eager to play their role in the fight against the 'baroni' (big shots) of the university, the old professors who were running CNR and its network of institutes and research centers. In May also the central building of CNR was occupied. INAC researchers, including myself, took an active part in the initiative. A few months earlier I had been elected as a delegate of INAC to the Council of the Union of CNR researchers, and this position had allowed me to understand the overall organization of CNR and to get in touch with other groups of CNR researchers who in various institutes of

---

[12] AICA was founded in 1961. From 1961 to 1967 the President of AICA was Aldo Ghizzetti, the Director of INAC.

mathematics and physics were working in the embryonic world of computer science.

In the gatherings and discussions that took place during the occupation, the conflict between applied and pure sciences started to emerge among the researchers. In particular those of us working in the domain of computer science added to the general issues of the protest of CNR researchers our particular feelings of frustration about the inadequate support that the new discipline was getting from CNR governing bodies, namely from the Mathematics Consulting Committee on which INAC and other computer science initiatives depended. In fact mathematicians at that time viewed computers just as tools for numerical computing, totally ignoring that, behind the instrument, another new mathematical discipline, the science of computing, was being shaped. This was not only happening in Italy but was a general attitude of mathematicians toward computer science.

One of the leading figures in theoretical computer science, Michael Rabin, recalls[13]: "There was absolutely no appreciation of the work on the issues of computing. Mathematicians did not recognize the emerging new field", and from this point of view it is also interesting what Edsger Dijkstra said about his appointment at the Department of Mathematics at Eindhoven University of Technology[14]: "Later I learned that I had been the department's third choice, after two numerical analysts had turned the invitation down". Still we felt that in Italy the situation was somewhat worse than in other European countries and for this reason we were fighting to gain a recognition of our discipline from CNR.

The occupation of CNR went on for about one month. In this period we had frequent exchanges of ideas with other branches of the student movement. I remember meeting a large variety of people, most of them young scientists interested in their research work but also excited about the exceptional political season that universities were experiencing. Among others I also happened to meet extreme left 'comrades', who used to cite the words of Lenin in each one of their sentences and who were also teaching other students how to prepare 'Molotov cocktails' in view of future possible clashes against the police! 'Amazing those years', to phrase it with the title of a book recently written by one of the leaders of the Italian student movement.

Finally, at the beginning of June, the occupation of CNR ended peacefully. A big rally was organized during which the student movement presented a report of the OECD on the state of scientific research in Italy to researchers, university professors and the press. The thick (200 pages) report carefully analyzed all aspects of scientific research carried on in universities and in public research institutions (such as CNR, CNEN, INFN etc.) putting in evidence that the investment in research and education in Italy was lagging behind the other industrial countries and pointing out the deficiencies of the

---

[13] http://wikicontents.altervista.org/?q=Michael_O._Rabin

[14] *EWD1166: From my Life* in *People and Ideas in Theoretical Computer Science*, C.S. Calude Ed., Springer 1998

country in this field. In particular it was shown that Italy was only spending 0.68% of GDP on scientific research, much less than most Western countries. The report, not publicized by the government, fell into the hands of the leaders of the student movement who printed it and distributed it to the Italian press[15].

After the conclusion of the occupation my life and the life of the other INAC researchers went back to normal. I resumed my work on the evaluation of λ-expressions and I continued my collaboration with Alfonso Miola on our algebraic manipulation system.

In August 1968 an important conference took place in Edinburgh: the IFIP World Computer Congress. IFIP was at that time the unique international organization in which there was a joint participation of both the Western countries and the countries of the Eastern block (Soviet Union and the so-called 'satellites'). For this reason the world congress, taking place every three years, was a very important event in which the main novelties in the field were communicated to a very large audience. At the 1968 Congress the main event was the presentation and discussion of a new advanced programming language designed by the IFIP Working Group 2.1, Algol 68. Several scientists specialized in the design of programming languages had taken part in the effort. In particular an important contribution had been made by Adriaan van Wijngaarden and the language had been defined using his two level grammars. The presentation of Algol 68 was also the occasion for a comparison with other important, recently developed programming languages such as Simula 67 and, especially, PL/I, a language conceived by IBM with the aim to make it a 'universal' language both in terms of applications (that is capable of serving applications in scientific computing as well as in commercial business) and in terms of target architectures (to become the standard programming language for the family of computers IBM System/360). Unfortunately due to the complexity of their implementation neither Algol 68 nor PL/I (although used until recently) have ever become popular.

In any case the big star of the conference (or at least the scientist that most impressed me) was Jean Sammet, from IBM. Bravely standing in front of a (predominantly male) audience from all over the world she gave a very effective talk on the history of programming languages (a brief history after all, since most programming languages were less than 15 years old!). In the talk she gave an overview of the main directions present in the field, an overview that one year later was published in her most famous book, the 'tower of Babel' book *Programming Languages: History and Fundamentals*. The cover image, the tower of Babel, depicted in a very clear way the situation of confusion and bewilderment in the field at the end of the 1960s, an era in which most of the important programming paradigms had already been invented (at least in embryonic form: structured programming, object-

---

[15] Sad to say, since then the percentage of GDP devoted in Italy to Research and Development, after growing in the 1980s has dropped again and in 2015 was still at the level of a modest 1.33. (http://www.istat.it/it/archivio/ricerca+e+sviluppo)

oriented programming, functional programming, concurrent programming)
but users were not yet mature enough to understand the power of the different approaches and 90% of the applications were still happily running with
FORTRAN and COBOL.

Chapter 3
Counting Steps in Cory Hall

In 1968 Corrado Böhm was appointed full professor and left Rome. After a short period spent in Modena (where he taught numerical analysis) he moved to the University of Turin. It was time for me to prepare to find a visiting position in some institution abroad. During his trip in the US Corrado had established several interesting contacts and he presented to me various options. In particular he spoke to me with enthusiasm about a young researcher who had just moved to the University of California at Berkeley and who was working in the field of computational complexity, Manuel Blum. The scientific interests of Corrado were not limited to λ-calculus and I loved his ability to catch new important ideas in all areas of theoretical computer science. So I was rapidly convinced that his intuition about Blum's research work was correct. I wrote to Manuel who warmly replied to me offering hospitality at the Department of Electrical Engineering in Cory Hall. Then I applied for a CNR foreign scholarship.

At the beginning of February 1969, together with my wife, who did not feel confident to go on a transcontinental flight, I took a passenger ship, the ocean liner Cristoforo Colombo, in Naples, destination New York. The trip took six days. We spent our time on board in the most leisurely way: eating gorgeous meals, watching movies, dancing, playing games (that was the first and last time in my life I did skeet shooting!). On a foggy morning we finally found ourselves in front of the forest coast of Canada, and after a stop in Halifax we arrived in New York. On the evening of the next day we landed at San Francisco airport where Peter Deutsch had come to pick us up by car in order to take us to Berkeley.

The hotel where we spent the first few days was on Telegraph Avenue, at that time the heart of the hippy district of Berkeley, a few hundred meters from the campus of the University of California. With the help of the International House we easily found an apartment to rent and a few days later we were happily settled. The apartment was in Grove Street (renamed Martin Luther King Way in 1984) a long street that stretched through Berkeley

© Springer International Publishing AG, part of Springer Nature 2018

G. Ausiello, *The Making of a New Science*,

https://doi.org/10.1007/978-3-319-62680-2_3

from Albany in the north to Oakland in the south. We were living close to the Albany boundary but still in walking distance from the campus.

Meeting Manuel Blum was a great experience. A few years older than me but already with the aspect of a guru, with a mild hippy look: long hair, bell-bottom pants, speaking very calmly with a soft voice. He had the ability to phrase a difficult new problem in simple terms, abstracting the essence of it. His office door was always open and this allowed a continuous interaction.

Fig. 3.1 Manuel, Lenore and Avrim Blum in the early 1970s

Manuel Blum (Fig. 3.1) is yet another gigantic figure in theoretical computer science. He was born in Caracas in 1938 from European parents and he received his Ph.D. in mathematics at MIT in 1964. His research interests span from neurophysiology to recursion theory but his most renowned results are in the field of cryptography. In 1995 he received the Turing Award with the following citation "In recognition of his contributions to the foundations of computational complexity and its application to cryptography and program checking"[1].

In 1969 Manuel Blum was one of the leading figures in computational complexity, another domain of computer science that had just recently started to be explored and that would rapidly become one of the major research directions in our discipline. In computational complexity the main objectives of research are, on one side, the analysis of the computational cost (according to some measure such as number of steps or number of memory cells) required to run an algorithm for solving a given problem and, on the other side, the characterization of the intrinsic cost which any algorithm for solving the problem would incur (that is the complexity lower bound of the problem).

The following are some of the fundamental questions that people interested in computational complexity wanted to answer: given two different functions, which one is, in general, easier to compute? How do we measure the complexity of a function? How can we classify computable functions in terms of the resources (in particular time and space) we need to carry out their computa-

---

[1] It is interesting to observe that three of Blum's former students (Leonard Adleman, Shafi Goldwasser and Silvio Micali) are also recipients of Turing Awards.

tion? As simple as these questions may appear, just to state them properly we need a deep knowledge of the concepts of computation resources and of complexity measure. In the 1960s most concepts were still to be precisely defined and the field was starting to take the first steps.

The idea to analyze the efficiency of the solution of a given problem was, of course, already in the minds of the mathematicians and engineers who in the 1950s wrote the first computer programs. It was certainly in the mind of Edsger Dijkstra when he conceived in 1956 his algorithm for finding a shortest path between two nodes in a weighted graph, or in the mind of Ford and Fulkerson when, in the same year, they defined their algorithm to compute the maximum flow (and, as they showed, the minimum cut) in a network[2]. Also the fact that some problems were easy to solve and some were difficult was intuitively very clear, especially in those cases in which the search space of solutions exploded in an exponential manner. But the intuition was not enough to explain what exactly "easy" or "hard" meant and how one might prove the hardness of a problem.

In a "digression", contained in a famous paper published in 1965, *Paths, Trees and Flowers*, Jack Edmonds claims that "easy" means that "there exists an algorithm whose difficulty increases only algebraically with the size" of the problem, but he also admits that "since one is customarily concerned with existence, convergence, finiteness, and so forth, one is not inclined to take seriously the question of the existence of a *better-than-finite* algorithm". In other words, at that time, mathematicians were still interested in the finiteness of a problem and the matter concerning the existence of efficient algorithms for its solution was still regarded as uninteresting.

As happens sometimes in those lucky moments in the history of science, when the convergence of different research threads creates the conditions for a new leap forward, from 1960 to 1965 various lines of research converged to shape the theory of computational complexity.

The first thread was coming from formal language theory. The notion of a complexity classification of languages was indeed already implied by the Chomsky hierarchy, based on more and more powerful grammars and on the corresponding classes of languages (regular languages, context-free languages, context-sensitive languages, unrestricted languages). In parallel to the hierarchy of languages, at the beginning of the 1960s computer scientists had started to define, as a counterpart, a hierarchy of acceptors, abstract machines able to recognize the words belonging to such languages, level by level endowed with growing computational power. In this hierarchy the finite state automata, machines making use only of a finite amount of memory, had been proved by Kleene and by Rabin and Scott to be able to recognize exactly regular languages; the (non-deterministic) push-down automata, introduced by Oettinger in 1961, which could use an arbitrary amount of memory, but

---

[2] The notion of the relevance of computation cost is also raised in a famous letter that Kurt Gödel sent to John von Neumann in 1956 in which Gödel wonders what may be the cost of proving a logical assertion in relation to the length of the assertion itself.

only as a push-down store, had been proved to be able to recognize exactly the context-free languages by Schützenberger in 1963; the (non-deterministic) linear bounded automata, introduced by John Myhill in 1960, a particular type of resource-bounded Turing machine, that could only use a number of tape cells bounded by a linear function of the size of the input, had been shown a few years later, by Sige-Yuki Kuroda, to be able to recognize exactly the context-sensitive languages. For the recognition of unrestricted languages the full power of a Turing machine was required.

A second approach to the notion of complexity of computation was deriving from the studies concerning sub-recursive classes of functions, that is classes of functions that were less general than the whole class of computable (or recursive) functions and whose members could be computed by means of some kind of 'bounded' algorithm (e.g. algorithms using bounded primitive recursion or using a bounded number of nested iterations). On such basis various mathematicians (among them Grzegorczyk, in 1953, and then Axt and Cleave in 1963) had defined hierarchies of classes of functions of increasing complexity. Along the same direction were the results of Robert Ritchie, who in his PhD thesis (written in 1961 under the advisorship of Alonzo Church and published in 1963) defined a hierarchy of classes of functions where the functions of each class had a time complexity bounded by a function of the preceding class, and those of Dennis Ritchie who, in collaboration with Albert Meyer[3], introduced a hierarchy based on the depth of nesting of iteration in a simple programming language[4].

Another contribution that laid the foundation of computational complexity was provided by the work of Michael Rabin[5] appeared in 1960 (in an unpublished report whose title was *Degree of Difficulty of Computing a Function and a Partial Ordering of Recursive Sets*), in which the author introduced for the first time the notion of step-counting function (that is a function that, given a Turing machine M, returns, for any input x, how many steps M makes when it is initiated on x) and proved, by means of a diagonalization technique (that is the same technique used by Cantor to prove that the cardinality of the reals is larger than the cardinality of the natural numbers

---

[3] Albert R. Meyer has a fundamental role in the development of computational complexity theory. After working at Carnegie Mellon for a few years he moved to MIT in 1969 and he took his PhD from Harvard University in 1972 under the supervision of Patrick Fisher. Among his PhD students are some excellent scientists such as Nancy Lynch, Leonid Levin, Charles Rackoff, Larry Stockmeyer, and David Harel.

[4] Dennis Ritchie (1941–2011) is actually better known to the broader public for the implementation of the UNIX operating system (for which he received the Turing Award in 1983 together with Ken Thompson) and for having invented the C programming language, with Ken Thompson and Brian Kernighan.

[5] The name of Michael Rabin has occurred already and will repeatedly appear in these notes. Rabin (born in Wroclaw in 1931) is one of the gigantic figures who built the science of computing. He has been professor at various institutions, including the Hebrew University of Jerusalem, MIT and Harvard. He received the Turing Award in 1976 jointly with Dana Scott.

and subsequently used by Turing to prove the existence of non-computable functions), that there are arbitrarily difficult decidable sets.

Finally, the work based on the attempt to consider more realistic models of computation, such as for example the Register Machines, a simple abstraction of the von Neumann computer model, which (with its organization based on control and processing unit and memory store) is at the basis of real computer architectures, and to compare their power to the Turing machines, determined further advancements. In fact, results concerning the simulation of register machines with Turing machines and also the results concerning the simulation of multi-tape Turing machines with one-tape Turing machines showed that such simulations could be carried out effectively: in particular, in a time polynomial in the number of steps executed in the simulated model. This opened a window on the possibility to establish complexity properties that were independent from the machine models and led researchers such as Alan Cobham to postulate (in a statement that was subsequently called 'Cobham's Thesis', in analogy with Church's Thesis) that the class of 'easy' sets could indeed be identified with the class of sets decidable in polynomial time with a Turing machine, providing in this way a formalization of Edmonds' claim mentioned above.

It is in this culture medium that a major step forward took place. In 1965 two researchers at General Electric Research Labs in Schenectady, Juris Hartmanis[6] and Richard Stearns, published *On the Computational Complexity of Algorithms*, a seminal paper that marked the beginning of a research field and that now counts more than 900 citations. Analyzing the relevance and the impact of the paper (whose content had already been announced at a conference in 1964) Richard Stearns, in his paper *Juris Hartmanis: The Beginnings of Computational Complexity*, says: "Perhaps our paper could be characterized as the first direct assault on the structure of deterministic time complexity. In any event, the paper uncovered the basic properties of deterministic time complexity which are so familiar today. But in addition to the results per se, the paper contained definitions and a point of view which fit very well with the more general needs of the computer science community".

In the paper, the authors provided a formal precise definition of the concept of time complexity class, that is the set of all languages that could be recognized by a (deterministic) Turing machine in a number of steps bounded by a given computable function $t(n)$ (n being the length of the input string, the description of the problem instance), and proved several fundamental results. Among them, the most interesting (still today the main result that allows the separation of complexity classes, that is proving that two classes do not coincide) was the time hierarchy theorem that states that a slight increase in the time provided to carry out a Turing machine computation allows the recognition of new languages, that could not be recognized within the previous time bound.

---

[6] Juris Hartmanis has been a leading figure of the field of computational complexity since the early 1960s. We will find references to his name and his work repeatedly in these notes.

For example, if t is a function that satisfies some reasonable properties, raising the computation bound (the number of steps allowed to be executed by any Turing machine) from $t(n)$ to a bound larger than $t(n) \log n$, say to $t(n) \log^2 n$, is enough to allow the solution of new problems that cannot be solved within the bound $t(n)$. As a consequence of this work the concepts of complexity upper bound and lower bound could be precisely defined. Various other important results regarding computational complexity would follow in the subsequent years, such as the results of Hartmanis, Lewis and Stearns on the space hierarchy and the most impressive result by Walter Savitch regarding the space bounded simulation of non-deterministic Turing machines by means of deterministic machines.

Beyond the specific results it is important to underline that the main achievements of the paper were of methodological nature. When he recalls those days, Juris Hartmanis says in his Turing Lecture: "To capture the quantitative behaviour of the computing effort and to classify computations by their intrinsic computational complexity, which we were seeking, we needed a robust computing model and an intuitively satisfying classification of the complexity of problems. The Turing machine was ideally suited for the computer model, and we modified it to the multi-tape version. To classify computations (or problems) we introduced the key concept of a complexity class in terms of the Turing machines with bounded computational resources".

Fig. 3.2 Juris Hartmanis (left) and Richard Stearns at the time of their joint work

For their work, Hartmanis and Stearns (Fig. 3.2) received the Turing Award in 1993 "in recognition of their seminal paper which established the foundations for the field of computational complexity theory". One of the founding fathers of this research domain, Stephen Cook (who received the Turing Award in 1982), says that Hartmanis and Stearns' paper, "was widely read and gave the field its title". The relevance of the work is clearly acknowledged also in the words of another Turing Award winner, Richard Karp, who in his Turing Lecture in 1985 said: " ... but it is the 1965 paper by Juris Hartmanis and Richard Stearns that marks the beginning of the modern era of complexity theory. Using the Turing machine as their model of an abstract computer, Hartmanis and Stearns provided a precise definition of the 'complexity class' consisting of all problems solvable in a number of steps bounded by some given function of the input length. Adapting the diagonalization technique that Turing had used to prove the undecidability of the Halting Problem, they proved many interesting results about the structure of complexity classes. All of us who read their paper could not fail to realize that we now had a satisfactory formal framework for pursuing the questions that Edmonds had raised earlier in an intuitive fashion, questions about whether, for instance, the traveling salesman problem is solvable in polynomial time."

One remarkable and long-lasting aspect of Hartmanis and Stearns' work was that Turing machines were adopted as the suitable computation model for analyzing the execution cost of algorithms. The relevance of this choice is expressed by John Hopcroft who in his own Turing Lecture (Hopcroft received the Turing Award in 1986) says: "Turing's work might have remained in the realm of mathematics and logic were it not for a seminal paper on computational complexity of algorithms by mathematicians Hartmanis and Stearns. They measured the complexity of an algorithm by the number of steps needed for its execution and used this method to develop a theory of complexity classes".

When all this happened Manuel Blum was still a student at MIT. In his PhD thesis, written under the supervision of Marvin Minsky, he started to develop an abstract approach to computational complexity with the aim of investigating what are the essential properties common to all complexity measures. His work had been strongly influenced on one side by Hartmanis' work on Turing machine computations and, on the other side, by the 'machine-independent' approach used by Hartley Rogers in his lectures on recursion theory at MIT. In this approach, instead of considering a specific machine model (e.g. Turing machines), computability theory was developed for any possible 'effective enumerations of partial recursive functions' corresponding to any possible universal machine model.

Manuel Blum, building on Rabin's work, had the idea to extend the same approach to step-counting functions (or if you like to all possible computation resources: time, memory etc.), assuming that in order to be an acceptable complexity measure, a step-counting function should satisfy two natural axioms: first it should be defined if and only if the computation of the corresponding machine halts and, second, we should always be able to decide whether a given number of steps (a given amount of resource) is enough to carry out a halting computation. Using this very abstract point of view Blum was immediately able to prove very general results that would hold for all complexity measures. For example he was able to show that all complexity measures are recursively related, that Rabin's result on the existence of arbitrarily difficult sets would hold for any complexity measure, and several other, more involved, results like the speed-up theorem and the compression theorem which generalized to all complexity measures results known for Turing machines' time and tape measures.

His paper *A Machine-Independent Theory of the Complexity of Recursive Functions* (which has gathered until now more than 900 citations), written in 1965 and published in 1967, soon prompted other studies, and when I arrived in Berkeley abstract computational complexity was one of the main research directions in the field. The impact of Blum's work in the theoretical computer science community became clear to me when in May 1969 I attended the first *ACM Symposium on Theory of Computing (STOC)*. In 1968 ACM had started the *Special Interest Group on Automata and Computability Theory (SIGACT)* and Patrick Fisher[7], the first Chair of SIGACT, at that time a professor at Waterloo, had created an annual conference devoted to the theory of computing that was held for the first time in 1969 in Marina del Rey, Los Angeles. STOC would become one of the most important world conferences in the field. It was the second conference entirely devoted to theoretical computer science, the oldest being the *IEEE Annual Symposium on Switching Circuit Theory and Logical Design* that had been created in 1959 and subsequently changed focus covering other domains such as automata, formal languages, algorithms and other topics in theory (in 1966 the name of this conference was changed to *Switching and Automata Theory* and later, in 1975, the name again changed to *Foundations of Computer Science, FOCS*). Since the early 1970s STOC and FOCS have been the major world conferences in theory of computing.

So, at the beginning of May I went to Los Angeles. Attending STOC was a great occasion to meet with some of the leading figures of the field and to see what were the main research directions (see Appendix A). Formal languages and automata were still the major domains. Out of 31 papers presented at the conference, 15 were devoted to formal languages and automata. Most papers addressed regular languages, context-free languages, abstract families of languages, push-down automata, and major attention was also devoted to applied topics such as syntax analysis for special classes of languages (precedence languages and $LR(1)$ languages). But the impact of Blum's work on the theory community was evident. Five papers could be classified as abstract computational complexity papers, that were produced by three major groups: Blum himself, Juris Hartmanis' students at Cornell and Albert Meyer's students at Carnegie Mellon. In particular, remarkable papers were presented by Allan Borodin, who showed that in any complexity hierarchy there are arbitrarily large gaps, and by Ed McCreight[8], who showed that the infinite union of complexity classes is still a complexity class, implying that, for example, there exists a computable function r such that the primi-

---

[7] Patrick Fischer (1935–2011) was one of the most important scientists who laid the foundations of computational complexity theory. He had been student of Hartley Rogers at MIT and among his students at Harvard are some prominent scientists such as Albert Meyer (Fig. 3.3), Dennis Ritchie and Arnold Rosenberg.

[8] Indeed, later, Ed McCreight became much more famous for having invented in 1972, in conjunction with Rudolf Bayer, B-trees, one of the most important physical data organization in data bases.

tive recursive functions are exactly the functions that are computable within the time bound r. Finally at the conference a gem of the theory of computing was presented: Savitch's result, stating that a language recognized by a non-deterministic space-bounded Turing machine that makes use of $s(n)$ tape cells can be recognized by a deterministic machine that makes use of $s^2(n)$ tape cells (provided $s(n) \geq \log n$). The paper, only two pages long, is a landmark in the history of computer science, essentially the only non-trivial result known concerning the simulation of non-deterministic computations by means of deterministic ones[9].

Fig. 3.3 Albert R. Meyer

The conference in the Los Angeles area was also for me a chance to visit southern California. I boldly decided to cross Death Valley with my car, a very old Mercedes, but the excursion was rather short: the car stopped in Emigrant Canyon with the water tank completely empty, fortunately not too far from a gas station. When I went back to Berkeley I sold the Mercedes and I bought a blue Ford Mustang. Then I asked Manuel to assign me some research problems to tackle.

Of course, Berkeley did not mean only Manuel Blum. Considering the three departments that in those years were active in theoretical computer science, in computability, in mathematical logic (Electrical Engineering, Computer Science[10] and Mathematics) the list of names of professors giving graduate courses in these areas at UC Berkeley was impressive and their lectures were at the forefront of the discipline. Let us see how Richard Karp describes the scientific environment in Berkeley, where he had moved in 1968: "My

---

[9] The journal version of this paper appeared in *Journal of Computer and System Sciences* in 1970 and now counts more than 1200 citations. In the paper Walter Savitch gives credit to Lewis, Hartmanis and Stearns for having introduced the same 'divide and conquer' technique in the proof that all context-free languages could be accepted in space $log^2(n)$ in their paper *Memory Bounds for Recognition of Context Free and Context Sensitive Languages* published a few years before.

[10] In 1968 a group of Electrical Engineering faculty members transferred to the College of Letters and Science to participate in establishing a Department of Computer Science. A 1973 merger formed the Department of Electrical Engineering and Computer Sciences, greatly broadening the scope of education and research activity (http://www.eecs.berkeley.edu/department/history).

new circle of colleagues included Michael Harrison, a distinguished language theorist who had recruited me to Berkeley, Eugene Lawler, an expert on combinatorial optimization, Manuel Blum, a founder of complexity theory who has gone on to do outstanding work at the interface between number theory and cryptography, and Stephen Cook, whose work in complexity theory was to influence me so greatly a few years later. In the mathematics department, there were Julia Robinson, whose work on Hilbert's Tenth Problem was soon to bear fruit, Robert Solovay, a famous logician who later discovered an important randomized algorithm for testing whether a number is prime, and Steve Smale, whose ground-breaking work on the probabilistic analysis of linear programming algorithms was to influence me some years later".

Although I was not formally a PhD student (my official status was 'research associate') the first thing I asked Manuel was to help me to establish a graduate education program and also to choose a few fundamental undergraduate courses that I might take in the Spring and Fall Quarters of 1969 and in the Winter Quarter of 1970, to fill the gaps I had in my education in logic and discrete mathematics. As auditor in most cases I succeeded to agree with the lecturers that they would also correct my homework. An outline of the content of the lectures may give an idea of what topics were considered hot fundamental topics in theoretical computer science education in that period and may illustrate what an exceptional observatory over the new frontier of theory UC Berkeley was at that time.

Among the first courses I attended were naturally Manuel Blum's courses in *Automata Theory* and *Theory of Algorithms* (A and B). The automata theory course was addressing various kinds of automata (McCulloch and Pitts models of neurons, Mealy and Moore machines), and was partly based on the famous 1959 paper by Rabin and Scott, *Finite Automata and Their Decision Problems*,[11] concerning one-way and two-way automata, deterministic and non-deterministic automata and regular expressions. Finally in the course various problems regarding automata were also considered (diagnosis problem, homing problem, identification problem etc.).

The theory of algorithms program (addressing computability and complexity) was divided into two parts. In part A recursive function theory was presented following Rogers' book *Theory of Recursive Functions and Effective Computability*[12], a new intuitive and appealing approach to computability based on proofs that were, so to say, carried out by Church's Thesis. In other words the recursive functions used in the proof of advanced theorems (e.g. Kleene's recursion theorem or Rice's theorem) were defined informally, with-

---

[11] For this paper "which introduced the idea of nondeterministic machines, which has proved to be an enormously valuable concept" Dana Scott and Michael O. Rabin obtained the Turing Award in 1976.

[12] Rogers' book, appeared in 1967, had a very strong influence on the research work that was at the time carried on in computability and in complexity theory. Richard Karp, who had been Roger's student at MIT, in his Turing Lecture expresses his admiration for the role that the book had and defines it a "superb book".

out requiring the construction of a detailed machinery, but with just enough formalism to grant their 'effectiveness' thanks to Church's Thesis. The same approach was used in the part of the course in which Blum presented his results in abstract computational complexity theory.

The course of *Theory of Algorithms* B was instead quite classical: construction of specific Turing machines, universal Turing machines and noncomputable functions. In any case listening to Manuel Blum's lectures was a great pleasure. His lectures were inspiring and he had the ability to involve the students just with his natural way to introduce problems and to present the ideas for their solution. Thirty years later, in a paper written in 1998[13], his colleague Richard Karp says: "A major reason for the success of theory at Berkeley has been Manuel Blum, a deep researcher, a charismatic teacher, and the best research adviser in all of computer science".

In Winter 1970 I took another fundamental course: Stephen Cook's course on Random Access Machines. Stephen Cook (Fig. 3.4) had received his PhD from Harvard University in 1966 and soon after he had moved to Berkeley where he stayed until 1970. Then he went to Toronto where he is still now. His aspect appeared to me rather cool and stiff, so different from the warm outgoing character of Manuel Blum, but his lectures were driving the students right into the heart of computational complexity, where all threads I mentioned before were crossing.

The aim of the course was to analyze the complexity of computing functions by making use of a computation model more realistic than Turing machines, so the course started with a presentation of 'random access machines' (RAMs, also known as 'register machines'), a model inspired by the von Neumann model of real computers. Cook explained to us how to analyze the complexity of RAM programs. For this part of the course he used a recent

Fig. 3.4 Stephen A. (Steve) Cook

---

[13] *The Mysteries of Algorithms*, in *People & Ideas in Theoretical Computer Science*, C.S. Calude Ed., Springer 1998.

book, which appeared in 1967, Minsky's *Computation. Finite and Infinite Machines*. The course also covered cornerstone material in computational complexity such as the time and space hierarchy theorems of Hartmanis, Lewis and Stearns, but the most important part was to show the polynomial-time simulations between RAMs and Turing machines that led to Cobham's thesis. Another part of the lectures concerned formal languages (in particular context-free languages and push-down automata) and were based on the brand new book by Hopcroft and Ullman, *Formal Languages and Their Relation to Automata*, that for decades has been the fundamental reference book covering the Chomsky hierarchy of languages and the corresponding hierarchy of automata.

As I said Richard Karp (Fig. 3.5) had also recently arrived at Berkeley: born in 1935, he had received his PhD at Harvard in 1959 in applied mathematics and then started working at IBM Watson Research Center in Yorktown Heights. In 1968 he moved to UC Berkeley, where, as he says, he had been recruited by Michael Harrison and became professor of computer science and operations research. From his aspect and the way of speaking you could easily recognize an East Coast gentleman, a character that he has maintained even after living for such a long time in California. In his lectures Karp presented fundamental algorithms for searching and sorting. Most of them are now considered standard course material but at that time parts of the topics were still rather new. Quicksort had been invented in 1960, AVL-trees in 1962, Heapsort in 1964. B-trees, one of the most important data structure for information management, were not yet known in 1970.

Fig. 3.5 Richard M. (Dick) Karp

But the most interesting part of Karp's course was the design techniques for efficient algorithms such as Horner's rule for evaluation of polynomials and what was later called the 'divide and conquer' technique, that consists in decomposing a problem into subproblems and in reconstructing the solution of the original problem by suitably assembling the solutions of the subproblems. Beside the classical application in the mergesort algorithm Karp illustrated the application of the divide and conquer technique in Karatsuba's algorithm for fast multiplication of two n digit integers and in the brand new result concerning matrix multiplication in time $O(n^{2.81})$, published by Volker Strassen in 1969 (the result that for the first time showed that n^3 elementary multiplications are not necessary to multiply two $n \times n$ matrices). The course also included a seminar given by a PhD student, Ivan Havel, on the parallel computation of polynomials. Ivan was the brother of Václav Havel, the well known playwright, one of the leaders of the Prague Spring, who, after the fall of the Iron Curtain became first President of Czechoslovakia and, subsequently, President of the Czech Republic.

In Fall and in Winter I took Julia Robinson's courses in Metamathematics[14]. The first course addressed classical mathematical logic, number theory and model theory. The second course was recursion theory based on Kleene's equational approach. Kleene's book *Introduction to Metamathematics* had been one of the holy books that Corrado Böhm had given to me to study when I was still a student and now it was a pleasure for me to go again through those pages under the guidance of Julia Robinson. But in the second Robinson's course a brand new, exciting jewel was hidden: the proof that the integer solution of Diophantine equations is undecidable, that is the (negative) solution of Hilbert's Tenth Problem. Diophantine equations are equations of the type $a_1 x_1^p + a_2 x_2^q + \cdots + a_m x_m^t = 0$ where $a_1, \ldots, a_m$ and $p, q, \ldots, t$ are given integers and $x_1, \ldots, x_m$ are the variables. Given an equation of this type the question is whether there exist integer values for all the variables that satisfy the equation[15]. David Hilbert included the problem to find an algorithm to decide the existence of integer solutions for Diophantine equations in his list of 23 problems presented in 1900 as the big challenges for mathematics in the 20th century.

Despite several efforts and despite several partial results achieved, in particular, by Martin Davis, Hilary Putnam and Julia Robinson, toward proving the non-existence of this algorithm, that is the undecidability of the problem, the question had proved difficult and was still open at the beginning of 1970. In February 1970 a breakthrough was announced: a Russian researcher, Yuri Matijasevic, had filled the last gap that consisted in providing a set of Diophantine equations that allow integer solutions if and only if one of the variables assumes as values the Fibonacci numbers. This completed the proof that the integer solvability of Diophantine equations is undecidable. In US the news was brought by John McCarthy who on February 9th had assisted in a lecture given in Novosibirsk by a Russian mathematician. At the begin-

---

[14] Julia Hall Bowman Robinson (1919–1985) had obtained her PhD at Berkeley under the advisorship of Alfred Tarski. In 1975 she became full professor at Berkeley. All of her outstanding scientific life was devoted to the study of decision problems in mathematical theories and, especially, to Hilbert's tenth problem.

[15] For example the equation $x^2 + y^2 - z^2 = 0$ has the integer solutions $x = 4$, $y = 3$, $z = 5$. Note that the question concerning the existence of integer solutions for the particular class of Diophantine equations of the type: $x^n + y^n - z^n = 0$ is instead decidable and has been proved to have a negative answer for any $n > 2$ and $x, y, z \neq 0$ (a result that is also known as Fermat's Last Theorem).

ning of March Julia Robinson presented to us the full proof, based on the notes of this lecture.

Finally I had the chance to follow one of the first courses in Artificial Intelligence, taught by Bertram Raphael, former student of Marvin Minsky, who at that time was a researcher at the Stanford Research Institute – Artificial Intelligence Center (the cradle, together with MIT, of most work in AI in the US in those years) and was soon to become Director of the center. The course was very rich: programming languages for AI such as LISP and SNOBOL, theorem proving, question-answering systems, natural language processing. During the course Jim Slagle (another student of Marvin Minsky, who was already famous for his work on symbolic integration), presented also his work on heuristics for games, a central topic in AI, and taught us the algorithm A^*, a heuristic technique used for exploring the state space in games and problem solving, just recently invented by Hart, Nilsson and Raphael.

The atmosphere of the department was in general rather friendly and after a while I happened to know many students and faculty members and have nice discussions with them. In particular I remember Jim Gray[16] with his long blond hair and a colored beads necklace, who just got his PhD in 1969. I also remember Lotfi Zadeh, at that time already a senior professor (he was born in 1921 in Azerbaijan and had been hired at Berkeley as full professor in 1959), who was already famous for having introduced the notion of 'fuzziness' in computer science, one of the first attempts to model uncertainty in computing. His 1965 paper *Fuzzy Sets* on *Information and Control* is one of the most cited computer science papers ever (60,000 citations). Lotfi was so kind to invite me and my wife for dinner to his place and took several photos, one of his favorite hobbies.

Beside the regular courses various seminars from outstanding scientists were also frequently held in Cory Hall and in the Computer Science Department. UC Berkeley has always been a crossroad and all major scientists regularly visited to deliver their talks. Among them I remember following Kenneth Krohn who presented the elegant algebraic theory of machines that he had developed in 1962 with John Rhodes, and Donald Knuth (Fig. 3.6) who gave a beautiful lecture on 'analysis of algorithms', a new paradigm that was just starting to take root in the field[17]. Juris Hartmanis also came to give

---

[16] Throughout his career Jim Gray worked for several companies, IBM, Digital, Microsoft etc. In 1998 he received the Turing Award "for seminal contributions to database and transaction processing research and technical leadership in system implementation". In 2007 he disappeared in the waters in front of San Francisco where he had gone on a boat to scatter his mother's ashes.

[17] Donald Erwin (Don) Knuth (born in 1938) has been a professor at Stanford since 1968. His contributions to theoretical computer science have an immense value and range from algorithms and data structures to compiler design and to semantics. In particular he is known for the encyclopedic treatise *The Art of Computer Programming* whose first volume appeared in 1968 while Volume 4 was published in fascicles between 2005 and 2015. For "his major contributions to the analysis of algorithms and the design of programming languages" he was assigned the Turing Award in 1974.

Fig. 3.6 Donald Erwin (Don) Knuth

a talk, at the end of April 1969, on his way to the first STOC conference. At that time he was chairman of the recently founded Computer Science Department at Cornell University. A few days earlier Cornell had been the stage for a black demonstration: a group of black students of the Afro-American Society had taken over the student union building asking the university to start a black studies program. In a rally the students made the black power salute and some of them showed guns to enforce their request. Before starting his lecture (a beautiful lecture on some recent results of his students in the field of abstract computational complexity) Juris opened his jacket to show, with an ironic smile, that he was not carrying any weapon.

My life as a student in Berkeley went on for about one year. Beside taking classes and doing homework I was carrying on the research work suggested by Manuel. In particular I was working on a few subjects. Among them one line concerned the study of resource-bounded computations of universal functions, corresponding to the situation in which we run in parallel the computation of the universal function and the computation of the resource bound (like a clock). The study, which was accepted at the 4th Annual Princeton Conference on Information Science and Systems, was carried on both in the Turing model, in order to investigate the relationship between resource-bounded computations and sub-recursive hierarchies, and in the abstract complexity model, in order to verify how much the various complexity measures could influence the cost of a universal resource-bounded computation. A second line of research, whose results I recklessly but successfully submitted to the Second STOC conference, concerned the analysis, in the abstract computational model, of the cost of computing a function whose convergence was determined by the convergence of a set of other partial recursive functions. The third line, whose results ended up later in a journal, was based on the idea, suggested by Manuel, of weakening the axioms for complexity measures in order to capture more closely the properties of measures like computation space that may be finite also in the case that the computation does not halt but, in this case, it is possible to decide whether the computation is going to halt or not. The aim of the research was primarily to show that this weakening left valid all the main results developed until now in abstract computational complexity and, second, to analyze the recursion theoretic properties of the set of input values for which the step counting function of the universal function is defined (e.g. the space used by the computation of the universal function is finite).

In the evening I used to study and do my homework. The radio station KFRC was playing the Beatles' *Get Back*, the Jackson Five's *I Want You Back* and Bob Dylan's *Lay Lady Lay* (just recorded in the 1969 LP *Nashville Skyline*). A great KFRC program that I recorded on the portable recorder that had been given to me by my friend Lello Zaffiro, an Italian chemical engineer working at Cutter pharmaceutical laboratories, was the History of Rock'n Roll, 50 hours with a wonderful collection of the best of rock and rhythm and blues music from 1955 to 1970. On a few occasions I also had the chance to attend some concerts given by rock'n roll heroes, such as Fats Domino and Little Richard.

Sometimes, in the evening, I met up with other Italian students to play *Risiko* and *Scrabble* or just to chat about the troublesome political period that Italy was traversing; the news regarding terrorist attacks and bombings in Italy were making the headlines in US magazines. During the weekends we used to go to the university swimming pool at Strawberry Canyon and on Sunday mornings we attended the Holy Mass at Newman Hall. For us it was a new experience to see young boys and girls with guitars singing 'peace and love' songs such as *We Shall Overcome* or Pete Seeger's *To Everything There is a Season* during the mass and even songs from the musical *Hair* like *Let the Sun Shine in*.

Beside Italian students, in Berkeley it was possible to meet also various Italian UC professors such as the physicist and Nobel laureate Emilio Segré and the computer scientist Domenico Ferrari, and some Italian visiting professors, such as the mathematician Alessandro Figà Talamanca. A few times I went to Stanford to listen to some talks or to meet Italian colleagues who were visiting the university or SRI. On one of these occasions I met Ugo Montanari[18], starting a deep and long-lasting friendship.

Also the exciting life in Berkeley was a topic to which we, in the Italian community, devoted long discussions. As you may know in 1969 Berkeley was not a place where you could feel annoyed. Downtown, the streets (Telegraph Avenue in particular) were full of hippies: young men with long hair and colored shirts and young women with full skirts and smiling faces. At the corner young people were selling the militant newspaper *Berkeley Barb*. Every day, at noon, in Sproul Plaza, the south entrance of the UC Campus, activists of various political or religious movements addressed the students. The first

---

[18] Ugo Montanari is one of the leading figures in the Italian academy in the field of theoretical computer science. His scientific interests span from program semantics to artificial intelligence and to combinatorial optimization. He became a full professor in Pisa in 1975 and since then he has educated several generations of excellent scientists.

striking event was, in May 1969, the 'People's Park War'. The university wanted to transform a small abandoned site along Telegraph Avenue into a sport field but the residents and the hippy community had other plans and had already started to transform the site into a public park by planting trees, flowers and shrubs. The students and the Free Speech Movement activists joined and when the police cleared the site and installed a fence around it, the confrontation became inevitable. On May 15 the students taking part in a rally at Sproul Plaza started moving down Telegraph Avenue and the crowd, grown to about 6,000 people, entered into a violent clash with the police. The police reacted by spreading tear gas and apparently also firing buckshot. It was the hardest confrontation taking place in the streets of Berkeley.

In the same year the protest against the Vietnam War was also very active but this issue involved the whole Bay Area. In 1968 the Têt offensive had shown how powerful were the Viet Cong (despite the massive US bombings), and some terrible cases such as the massacre of My Lai had prompted a strong reaction not only of the peace movement but more generally of public opinion. In November 15, 1969 a big march against the war took place in San Francisco, at the Golden Gate Park, as in other cities in the US, on the occasion of the 'Moratorium to end war'.

At the beginning of 1970 I started to make plans to go back to Italy. I was supposed to take a ship in New York to return to Naples in May and I decided to leave Berkeley in March in order to spend a couple of months visiting some universities on the East Coast. In any case, at the beginning of May I should have been in Northampton, Massachusetts, for the STOC conference where I was supposed to present my paper. The trip was organized with the help of the International House who found, in the places where I wanted to stay, a family willing to host me and my wife for one or two weeks. I also started to write letters to the people I wanted to meet and all of them very kindly answered with an invitation to visit their departments. Beside other cities, the trip included Philadelphia, where I was invited by Hisao Yamada to give a talk at the Moore School of Engineering, Ithaca, where I was invited by Juris Hartmanis to visit his recently created Department of Computer Science, Boston, where I visited Joel Moses, the leader of algebraic symbolic manipulation at MIT.

Again, thanks to my first wife's phobia for airplanes, the trip was a very interesting occasion to experience other ways of traveling. First we crossed the United States by train, from Oakland to Chicago, during three full days in the snowy countryside of the North Central States. Afterwards we rode the legendary Greyhound buses for the rest of our trip. Also the people we met during the trip contributed to make this experience so fantastic. We had the chance to meet people of various social classes: middle class public employees in Philadelphia, junior high school teachers in Ithaca, a lawyer in Washington and even a bishop in Boston, father of the former Governor of Massachusetts Endicott Peabody. All of them were apparently happy to host a young Italian

couple and interested to hear about life in Italy. All of them liberals and all
of them fiercely against the Vietnam war.

Meeting with Juris Hartmanis at Cornell was definitely the most important experience in this trip, both for the pleasure to meet such an outstanding
figure and for the consequence this visit had on future developments. Born
in Latvia in 1928, after the war he moved to the US[19] and received his PhD
in Mathematics in 1955 at Caltech. From 1958 to 1965 he worked at General
Electric and afterwards he became a professor at Cornell University where he
founded the Department of Computer Science. The idea to found a Computer
Science Department at Cornell was primarily supported by Cornell faculty
members working in operations research (Richard Conway) and mathematics
(Anil Nerode and Robert Walker) who obtained a one million dollars grant
from the Sloane Foundation and chose the first three professors: Juris Hartmanis, Pat Fischer (both working in computational complexity) and Gerry
Salton (a computer scientist working in information retrieval). Juris became
Chairman. This was a great choice, as David Gries, one of the first professors
of the department, hired in 1969, acknowledged in his speech held on May
6, 2001 on the occasion of Juris' retirement[20]. David says: "Yes, indeed. It
was Juris, the first chair of CS, who gave it its vision and, in his own inim-
itable, informal, but extremely effective leadership style, lead the department
to become one of the great CS departments". At the same time computer science was still somewhat in the making with respect to other well-established
disciplines and this aspect is well caught by Allan Borodin (one of the first
of Hartmanis' PhD students, actually the first student to earn a PhD in
Computer Science at Cornell) who, in his paper *Juris Hartmanis: Building a
Department – Building a Discipline*, says: "Hartmanis was the Chairman of
(what was then) a very questionable and undefined discipline" (cf. Fig. 3.7).

Juris can definitely be considered the scientist who most consistently has
pursued the study of computational complexity, a domain in which he has
educated various dozens of PhD students. On the other side the scientific
curiosity of Juris Hartmanis has always spanned a much larger horizon, and
concerned the broad field of computing and its peculiarity. Let us see how he
describes this field in his Turing Lecture: "Computer science differs so basi-
cally from the other sciences that it has to be viewed as a new species among
the sciences, and it must be so understood. Computer science deals with information, its creation and processing, and with the systems that perform
it, much of which is not directly restrained and governed by physical laws.
Thus computer science is laying the foundations and developing the research
paradigms and scientific methods for the exploration of the world of information and intellectual processes that are not directly governed by physical

---

[19] As he recalls in his Turing lecture, before arriving in the US he stayed at a displaced
persons camp in Germany where he received "a superb high school education from Latvian
academics who conveyed their enthusiasm for knowledge, scholarship and particularly for
science", an enthusiasm that Juris has assimilated and that has never abandoned him.

[20] http://www.cs.cornell.edu/gries/banquets/jhretireparty/gries.html.

Fig. 3.7 Allan Borodin

laws. This is what sets it apart from the other sciences and what we vaguely perceived and found fascinating in our early exploration of computational complexity."

When I entered his office in April 1970 he very kindly invited me to sit in his green armchair. I was very embarrassed and that armchair seemed to me bigger than it really was. Although Juris' approach is always very friendly he is also sometimes stiff and his smile can be formal, a bit wry. After a pleasant conversation we moved to the seminar room where I presented my work on weak axioms for computational complexity measures. In the audience was Robert Constable, who had joined Cornell in 1968, and who, at that time, was working in abstract computational complexity. In cooperation with Allan Borodin (who had received his PhD at Cornell and had just moved to Toronto the year before) he was working on subrecursive formalisms. This research thread had been started a few years before by Manuel Blum. In 1967 in another memorable paper, *On the Size of Machines*, Blum had shown that even if most functions we compute are subrecursive, using subrecursive formalisms to define them may lead to programs with enormously larger size. Somewhat in contrast with this result, Constable and Borodin were able to show that using subrecursive languages does not lead, instead, to a substantial loss from the point of view of efficiency. At the end of my talk Juris raised his thumb in a sign of approval. I was thrilled.

The next week I went to Northampton, Massachusetts, to attend the Second STOC conference. The conference was held, according to an everlasting tradition, in an anonymous dull place, at the crossing of two highways. No chance to do anything else but attend all talks.

Formal languages and automata, including syntax analysis, was still the major research domain, but complexity was a close second, and five papers were still devoted to aspects of abstract computational complexity. Among the other complexity papers an important paper was Stephen Cook's on path systems and language recognition, which can now be considered school book material. From the results presented in that paper, Savitch's theorem concerning space-bounded non-deterministic Turing machines and Lewis, Stearns and Hartmanis' results on the space required for recognizing context-free languages could be naturally derived. Other important papers were delivered by various other big names of the theory of computing: Aho and Ullman, Zohar Manna, Arny Rosenberg, Ron Book, Bob Floyd, Raymond Reiter. On the last day, the last two papers were presented by two young French guys: Jean François Perrot and Maurice Nivat. The papers were following the algebraic approach to formal languages and automata inspired by the guru of the French school of formal languages, Marcel-Paul Schützenberger[21]. At that time I did not know yet what role the two (and in particular Maurice Nivat) would play in European theoretical computer science and also in my life.

Again, as in the year before, attending the conference I had the impression that a wide uncovered territory was exposed in front of my eyes. At least apparently the hunt for interesting topics that were just waiting to be explored and mathematically analyzed was open. In view of my return to Rome I was feeling full of enthusiasm. The last occasion that I found exciting, the moment I still remember very clearly, was when, during the break after my talk, Dick Karp approached me and said that Juris had spoken to him about my other work on weak axioms suggesting that I submit the paper to the *Journal of Computer and System Sciences* (JCSS), of which he was one of the editors. JCSS was one of the first journals devoted to theoretical issues in computer science; it had started in 1967 and was already considered very prestigious. I felt flattered. That would be my first submission to an international journal.

---

[21] Marcel-Paul (Marco) Schützenberger (1920–1996) was an extremely interesting French scientist; first, in 1948, he obtained a PhD in medicine, then, after working in South East Asia for a few years, in 1953 he obtained a PhD in mathematics. Among a variety of important mathematical results, his most relevant contributions (in part achieved in collaboration with Noam Chomsky) were in the fields of combinatorics, coding theory and formal languages. Besides, biological aspects of science were always in his realm of interest. In particular he was interested in combinatorial aspects of evolutionism. Most important French figures of theoretical computer science, such as Nivat and Perrot, were his students.

Chapter 4
'I hate numerical analysis'

In June 1970 I was back to Rome, in my office at INAC, whose name had in the meantime changed to IAC, losing the 'National' epithet. Returning back to Rome from Berkeley was not easy. The situation at IAC had changed while I was away.

First of all Corrado Böhm had left. The few colleagues pursuing scientific interests in theoretical issues, Jacopini, Venturini, Miola and I, remained without a formal leader and gathered into a small research group. Other groups could be identified in operations research (Giorgio Gallo and Bruno Simeoni, under the leadership of Enzo Aparo), in graphics (Goffredo Pieroni), in programming languages, etc. Finally, the group of numerical analysts, lead by Dino Dainelli, continued their traditional work in various directions of calculus and applied mathematics, which was now supported by a new UNIVAC computer.

In second place the research situation in computing in Rome had also changed. In the late 1960s the perception of the role of computers in science and in society had grown. The field of electronic computing was now called '*informatica*', a term coined in France[1] and rapidly adopted throughout Europe[2].

Initiatives devoted to education and research in informatics were arising in various universities (with the creation of the new degrees in *Scienze dell'Informazione*) and in the National Research Council. At the Faculty of Engineering in Rome Antonio Ruberti had created a new CNR research center (Research Center for Automatic Control and Computing Systems) and Paolo Ercoli had joined the center, together with most of the engineers and technicians who had been taking care of the INAC computing systems. The

---

[1] "Informatique: Science du traitement rationnel, notamment par machines automatiques, de l'information considérée comme le support des connaissances humaines et des communications, dans les domaines techniques, économiques et socials" (Académie Française, 1969).

[2] Apparently, in Germany the word *Informatik* was first used in 1968 by the federal government minister Stoltenberg, at the opening of a conference in Berlin.

© Springer International Publishing AG, part of Springer Nature 2018
G. Ausiello, *The Making of a New Science*,
https://doi.org/10.1007/978-3-319-62680-2_4

name of the new center reflected the ambitious idea of Antonio Ruberti that computing systems and control systems could be seen both as sub-domains of systems science, that is the discipline which studies complex systems, in terms of the time evolution of observed and measured system variables, possibly under different parameter configurations and initial conditions, an idea that affected the development of research in computer science in Italy for almost two decades.

Actually, this point of view was quite popular in the 1960s. In France, in 1967, a new research center had been created, IRIA, to develop in conjunction research efforts in the two domains of computer science and automation ('informatique et automatique'). The idea of a strong connection between computing and system sciences was also present in the title of some important international journals such as the *Journal of Computer and System Sciences*, *Information and Control* (now *Information and Computation*) and *Mathematical System Theory* (now *Theory of Computing Systems*). Despite their titles, which in some cases have been changed in order to reflect the scope of the journals in a better way, they are still today among the major journals in the field of theoretical computer science, and they never devoted much room to papers regarding system science.

In any case, the foundation of the new center had an important meaning and Ruberti's initiative could be considered meritorious: for the first time, there was a public research institution in Rome officially (although partially) devoted to computer science. Moreover, this institution was based in the Faculty of Engineering and was competing with IAC and with the traditional role that mathematicians used to have in this new domain.

The third important change concerned the director of the institute. The new director, who had replaced Aldo Ghizzetti, was Guido Stampacchia, an excellent and very cultivated mathematician, one of the European leaders in functional analysis. As one of the first steps in his role, he hired young and bright mathematicians working in the same field and did not seem very keen to develop a research direction in theoretical foundations of computer science. Indeed, he did not show much interest in applied mathematics, the original 'core business' of IAC, and for this reason after two years the Mathematics Consulting Committee of CNR decided to suspend him from office and to nominate first a commissioner (Michele Sce), with the aim of reorienting the institute in the direction of applied mathematics, and, after six months, a new director (Ilio Galligani), a numerical analyst.

In any case I started to wrap up my experience in the US and to write down the ideas I had developed during my trip. So I began to write the paper on weaker axioms that I was invited to submit to *JCSS* and at the same time I started to prepare a survey paper devoted to the various approaches to the notion of complexity in computer science. Besides, I was anxious to transfer to somebody the ideas and the excitement that I carried with me from the US and it was a very pleasant surprise when a colleague from Pisa, Antonio Grasselli, came to visit me in Rome with one of his brightest students,

Giuseppe Longo, to propose that I be the advisor of his master's thesis in mathematics[3]. I suggested to choose among some of the open issues in abstract computational complexity I had in mind and finally we decided to address the formalization in abstract terms of the amount of resources used by a machine during a computation.

The collaboration with Giuseppe was very fruitful and enjoyable, but the last months of 1970 were a tough period for me. On October 22 my father died of a heart attack at the age of 56 and six days later my first son, Gabriele, was born.

Beside my work in computational complexity, several other activities were carried on in our research group (whose official name was *Logic and Programming*). Marisa Venturini Zilli besides continuing to work in λ-calculus was initiating a successful research activity in automatic theorem proving and Puccio Jacopini was investigating λ-calculus models, following the approach indicated by Dana Scott in his breakthrough paper *Outline of a Mathematical Theory of Computation*[4]. At the same time Alfonso Miola and I went on with our collaboration on algebraic manipulation systems and more generally in languages, systems and applications for non-numerical computation. This research direction was, in a sense, our flagship activity and when the new IAC Director Ilio Galligani (an expert in numerical analysis) was appointed we welcomed him with a poster saying in capital letters: 'I HATE NUMERICAL ANALYSIS', the *motto* that was written on the back cover of all LISP reports printed at Bolt, Beranek and Newman.

Our diverging views on numerical versus non-numerical computing of course created a quarrel with the new director: the gap between our expectations on the development of theoretical computer science issues and the steering imposed on the institute by the Mathematics Committee toward research work in applied mathematics could not have been wider.

With the aim to find support for our ideas, our small group of theoreticians decided to establish contacts with colleagues in other CNR institutes and university research centers who were sharing with us the same interest in foundational topics in computer science: Naples, where Eduardo Caianiello, at the Institute of Theoretical Physics, had initiated a scientific school in cybernetics, the discipline aimed at studying and measuring information and its role in communication and control systems, and educated a large group of researchers who were now turning their interest to the new promising domain of computer science; Pisa, where the field of theoretical computer science had enormously grown with respect to what I had seen five years before, also thanks to the creation in 1969 of the new degree in Information Science; and Milan, where two institutions were competing in the field: the

---

[3] Giuseppe Longo, was a professor of computer science in Pisa until 1990 when he moved to the École Normale Supérieure in Paris where he is now *Directeur de Recherches*.

[4] In this paper, presented in 1970 at the *4th Annual Princeton Conference on Information Science and Systems*, Dana Scott laid the foundations of the so-called 'mathematical semantics' (after 1976 better known as 'denotational semantics').

Electronics Institute of the Polytechnic on one side and the Physics Institute
of the University on the other side.

Eduardo Caianiello's school of cybernetics is a legend in the Neapolitan
scientific scene and deserves a few words. Caianiello (Fig. 4.1) became full
professor of theoretical physics at Naples University in 1955. During the preceding years he had developed a strong interest in information theory and
cybernetics (in Wiener's words: "the scientific study of control and communication in the animal and the machine") thanks to the stimuli received by
Enrico Fermi, and he decided to start a research activity in this field in Naples.
The encounter with Valentino Braitenberg (a physiologist and neurologist)
encouraged him in this direction.

Soon he started to establish scientific contacts with Warren McCulloch and
Norbert Wiener, who subsequently visited the University of Naples several
times. In 1958 he organized a successful summer school in Varenna: although
officially devoted to information theory the school had indeed a larger scope
involving more broadly the whole area of cybernetics. Beside Wiener the
lecturers included pioneers of information theory such as Robert Fano and
David Huffman, linguists like Yehoshua Bar-Hillel and Morris Halle, mathematicians like Marcel-Paul Schützenberger, neurophysiologists like Braitenberg, etc. Among the students were Corrado Böhm and the future leaders of
the Italian school of system theory, Antonio Ruberti and Antonio Lepschy.
The tradition of Caianiello's schools continued to achieve strong success. The
schools were one of the ways in which he continued to gather a lively interdisciplinary community around his scientific project. After the school on
Information Theory, a school on *Cybernetics of Neural Processes* was organized in Naples in 1961, a school on *Automata Theory*, in Ravello, in 1964,
that we already mentioned as one of the landmarks in the early days of theoretical computer science in Italy, and finally a school on *Neural Networks* in
1967.

Fig. 4.1 Eduardo Caianiello

Let us see how Antonio Restivo[5], professor of computer science at the University of Palermo, one of the European leaders in formal languages and automata theory, describes the atmosphere that surrounded Caianiello's group to which he belonged after 1970. "An immediate effect of Caianiello's school has been to make Italian academic circles (still strongly attached to traditional disciplines) aware of the domain of cybernetics and to stimulate the attention of a large public toward this new research field. Among the great merits of Caianiello there has been the one to coagulate a large Italian scientific community in the field of information theory and cybernetics. This in particular happened thanks to the inherent interdisciplinarity of the field that has allowed the participation in research projects of scientists with eterogeneous education. The group created at Naples Institute of Theoretical Physics included physicists, chemists, mathematicians, logicians, engineers and biologists, and had collaborators also in the humanistic field such as, for example, experts in natural languages".

The activity of Caianiello continued with growing emphasis after CNR created the Center for Cybernetics at the University of Naples. The Center was organized in three Sections: the Theory group, the Laboratory of Neuroanatomy and Physiology, and the Laboratory of Electronics. The aim was as follows: first the Neuroanatomy Laboratory had to understand how the mind works, then the Theory group would define mathematical models, and finally the Electronics laboratory would take care of the physical implementation of the models into machines. This road map was indeed followed in some of the projects carried on in that period but the main difficulty was the lack of knowledge (that we still have today) about the way our mind works. Caianiello's ideas were indeed much too advanced with respect to the ability to understand the neural structure and behaviour of the brain, on one side, and, on the other side, with respect to the limited power of electronic devices available at that time.

Here is how Aldo de Luca (one of the first pupils of Caianiello, who later became professor at the University of Naples and was himself one of the European leaders in the field of formal languages) explains the downtrend of cybernetics[6]. "Essentially the program of cybernetics, although extremely interesting, was too ambitious and the results that were achieved were indeed disappointingly modest with respect to the expected goal. The program originally was to describe neural nets, to design mathematical models of the nervous system and from this to derive an explanation of high level mental functions such as learning, intelligence and even consciousness but the study and understanding of the central nervous system did not develop as expected ... Eventually from cybernetics a series of specific disciplines developed, each one with its own language and its own goals". Among them theoretical computer science.

---

[5] A. Restivo, *Gli esordi scientifici*, cit.

[6] A. de Luca, *Eduardo Caianiello e la nascita a Napoli di una ricerca interdisciplinare*, in *Memoria e progetto*, P. Greco e S. Termini eds., GEM 2010

In 1968 the center was transformed into a permanent structure, a laboratory, and was moved to Arco Felice, a beautiful area in the countryside, in the volcanic area of Pozzuoli, in front of the pleasant warm spring and health spa of Fratelli Damiani, but at that time the crisis of cybernetics was already evident. Although the laboratory was divided into sections (physics, neurophysiology, computer science) corresponding to the main components of the idealistic interdisciplinary concept of cybernetics, gradually, after the creation of the laboratory, each section started to follow the scientific trends considered more relevant in each domain. Neurophysiologists started to concentrate on the study of specific neurophysiological systems, such as, for example, the eye of the frog, physicists moved in the direction of solid state physics, computer scientists moved toward the new appealing subfields of theoretical computer science.

This was the situation when we approached the colleagues in Naples to analyze our chances of collaboration. In a series of meetings we discovered a very rich reality. Our colleagues, graduated in various southern universities between 1965 and 1970, and hired by the Laboratory of Cybernetics, were working in a wide variety of fields (cf. Fig. 4.2): algorithms for the analysis of natural language (Renato Capocelli[7]), automata theory (Antonio Restivo), codes (Aldo de Luca), fuzzy sets and uncertainty (Settimo Termini), models of computation based on rewriting systems (Giorgio Germano and Andrea Maggiolo). At the same time, the Cybernetics group that had remained at the Institute of Theoretical Physics was also very active: in particular, Francesco Lauria continued his research work in neural models, while Giuseppe Traut-

Fig. 4.2 Aldo de Luca (left) and Antonio Restivo

---

[7] Renato Capocelli, born in 1940, was one of the most brilliant Italian scientists in the field of Combinatorics and Information Theory. After working for a few years on the analysis of natural language in the Procrustes project under the guidance of Caianiello his interest moved to mathematical models in biology and then to the study of entropy and of information storage and compression. When he died at the age of 52 he was professor at 'Sapienza' University of Rome. All his friends always remember him for his incredible charge of enthusiasm and optimism.

teur and Leo Aloisio[8] turned their activity toward computability and complexity.

Then there was Pisa: in the early 1970s this city was to become the cornerstone of computer science in Italy. In Pisa there were several research institutions in computer science. The older one was CSCE (*Centro Studi sulle Calcolatrici Elettroniche*) that was later transformed into a research institute called IEI, *Istituto per l'Elaborazione dell'Informazione* (*Institute for Information Processing*). In the 1950s at CSCE the first attempts to construct an Italian computer had been made by Giovanbattista Gerace, giving rise to the CEP machine (Calcolatrice Elettronica Pisana) that had been inaugurated in 1961.

At the university the area of computer science had been strongly enhanced by the creation, in 1969, of the first degree (laurea) in computer science introduced in the Italian university system. The initiative was due to the vision of the Rector, the mathematician Alessandro Faedo, and to the energetic efforts of a group of brilliant professors, some working already at Pisa University, like Alfonso Caracciolo and Giovanbattista Gerace, and some recruited on purpose from universities in northern Italy, like Antonio Grasselli, Giorgio Levi, Fabrizio Luccio, Alberto Martelli and Ugo Montanari (Fig. 4.3). Originally the name assigned by the Ministry of Education to the new degree, whose creation was initially approved only for the Universities of Bari, Mi-

Fig. 4.3 Ugo Montanari (right) and Carlo Montangero

---

[8] Pantaleo (Leo) Aloisio, pupil of Caianiello, was an incredible character, always humorous and often ironic, expert of logic and of arts. Nobody was able to know how old he was. We only knew that he was teaching at the Mathematics Department of the University of Naples and that, despite not having a degree, he had a broad and deep knowledge of mathematical logic. Unfortunately he passed away in September 2011, leaving great sadness in all who knew him.

lano, Pisa and Torino, had been *Laurea in Scienze dell'Informazione*, and it was changed into *Laurea in Informatica* only about 20 years later. The university institute that was leading the program had the same name: *Istituto di Scienze dell'Informazione* (ISI).

Another remarkable institution devoted to computer science in Pisa was CNUCE, originally a computation center (literally: *Centro Nazionale Universitario di Calcolo Elettronico*) created in 1965 thanks to an agreement between the University elect of Pisa and IBM which donated a 7090 machine (replaced in 1970 by a 360/67 and subsequently by the most advanced machines of the 370 family). In the 1960s and in the 1970s CNUCE was the major CNR computing center, serving the needs of large scientific communities (mathematicians, physicists, chemists, engineers, but also linguists, musicians, physicians). When the Rector of Pisa University Alessandro Faedo became President of CNR in 1972, he had the idea to bring CNUCE under the CNR umbrella and CNUCE became an institute of CNR in 1974 maintaining at the beginning its mission of computing center and extending in this way a service to a larger community of users, including the entire network of CNR institutes.

In 1971, at the time in which the first contacts between IAC and Pisa researchers were established, all topics of computer science considered relevant in those years were addressed in these research institutions (IEI and ISI in particular): hardware design, operating systems, programming languages, software engineering, information retrieval, etc. Theoretical research issues were carried on by a fairly large group of people. In the section of IEI devoted to 'Non-numerical information processing' Antonio Grasselli, Giorgio Levi and Ugo Montanari were working on heuristic approaches to pattern recognition with applications to fingerprints and chromosome recognition. Besides, Giorgio Levi, Fabrizio Luccio, Alberto Martelli, Ugo Montanari were studying a variety of fundamental algorithms and data structures (hash tables, graph isomorphism, constraint networks, and-or graphs, etc.). Finally, Luigia Carlucci[9] (a former student of Alfonso Caracciolo) together with her husband Mario Aiello[10] and Gianfranco Prini were working on issues related to λ-calculus, LISP and the semantics of functional programming languages. Research on theoretical issues was also carried on in the section devoted to 'Programming systems', in which Alfonso Caracciolo, Renzo Sprugnoli and Alfio Andronico were applying Markov algorithms in the formal definition of the semantics of programming languages (such as LISP, Algol 60 and PL/1).

A third site where theoretical computer science studies were carried out was Milan. As I said above, in Milan two institutions were the most interesting. The oldest was the Computer Laboratory in the Electrical Engineering

---

[9] Luigia (Gigina) Carlucci Aiello (Fig. 4.4), born in 1946, would later become one of the greatest European leaders in the field of artificial intelligence. After having been a researcher at the Istituto di Elaborazione dell'Informazione in Pisa, in 1982 she became full professor at 'Sapienza' University of Rome. Her research interests have been mainly devoted to knowledge representation and automatic reasoning. She had important roles in

Fig. 4.4 Luigia (Gigina) Carlucci Aiello

Institute of the Polytechnic, lead by Luigi Dadda who was an electrical engineer, but nevertheless was very interested in developing computer science both in the applied direction and in the theoretical direction. The role of Milan Polytechnic in informatics has always been very important. In the 1950s the Center for Numerical Computing had been created at the Polytechnic and there, since 1954, was operating a CRC 102A, an NCR computer, the very first computer installed in Italy, a few months before the installation of the computer in Rome and the beginning of the research project that led in Pisa to the construction of CEP. Traditionally the major research area was in circuit theory (Luigi Dadda and Mariagiovanna Sami) but in the early 1970s new research directions were growing: databases and information structures (Giampio Bracchi), formal languages (Stefano Crespi Reghizzi, whose interest in the field of grammar inference, started at UCLA, had led to the presentation of a paper at the IFIP conference in 1971 in Ljubljana), software engineering (a new field in which a group of young and promising researchers was active: Carlo Ghezzi, Pierluigi Della Vigna, Dino Mandrioli) and artificial intelligence (Marco Somalvico[11]).

A second group was active in Milan, the Electronics and Cybernetics Group of the Physics Institute of the University lead by Gianni Degli Antoni[12] (Fig. 4.5), a young, extremely energetic professor with whom some very bright researchers were working on the most advanced theoretical subjects

_____

various international associations devoted to AI such as ECCAI and IJCAI and was also Chair of the IFIP Technical Committee for Artificial Intelligence.

[10] Mario Aiello, one of the most promising researchers of Pisa University died very young, in 1976, after a terrible illness that took his life in just a few months.

[11] Marco Somalvico was a pioneer of research in artificial intelligence and robotics and for three decades was one of the major Italian leaders in the field, until his premature death in 2002, at the age of 61.

[12] Gianni Degli Antoni (1935–2016) was a theoretical physicist and during his work in computer science he always maintained the cultural view, the openmindedness and the modeling attitude of a physicist. As a teacher Gianni did tremendous work in educating

Fig. 4.5 Gianni Degli Antoni

Fig. 4.6 Alberto Bertoni

of computer science (among them Alberto Bertoni[13] (Fig. 4.6), working in automata theory, Mario Ornaghi, working in grammar inference and Giorgio De Michelis working in programming languages).

Beside Rome, Naples and Milan, of course, theoretical computer science had bloomed in Turin, under the leadership of Corrado Böhm who had created the second *Corso di Laurea in Scienze dell'Informazione* and already raised various young promising scientists, in particular Mariangiola Dezani[14], Simona Ronchi della Rocca and Mario Coppo, all engaged in research in λ-calculus and in various issues related to information structures (cf. Fig. 4.7).

---

several generations of Italian computer scientists, always able to conjugate technological advances and theoretical insight.

[13] Alberto Bertoni (1946–2014) was one of the major Italian leaders in theoretical computer science. His incredibly vast culture, strongly influenced by his education in physics, spanned a wide range of subjects. In his obituary, which appeared in the journal Theoretical Computer Science, we can read: "The variety of his studies and research interests is really impressive: it includes computational complexity, computability, formal languages, counting and enumeration, probabilistic and quantum machines, neural networks and genetic models, learning algorithms with applications to bioinformatics".

[14] Mariangiola Dezani Ciancaglini, born in 1946, is a professor at Turin University. She is one of the most influential Italian theoretical computer scientists; in particular she had a fundamental role in the development of research on typed λ-calculus in Europe, an important subject with crucial applications in programming languages

Fig. 4.7 From left to right: Simona Ronchi della Rocca, Mariangiola Dezani, Mario Coppo

Finally a small but very active group (Renzo Pinzani, Giovanni Soda, Gianni Aguzzi) existed in Florence, working in logics and combinatorics.

Altogether, the research centers with whom we got in contact were eight. In the Spring of 1971 we organized several seminars and small workshops, alternating between the various sites, which led to a fruitful comparison of the research work and of the various approaches among the groups. As we will see such groups became the pillars on which in the subsequent years the first Italian research network in theoretical computer science was created.

Establishing such contacts was a very important step. Until then most of us had scientific partners abroad, most frequently in the States, and almost no contact with colleagues in other Italian institutions. Besides, it was important to prepare joint research projects to submit to CNR, to compare our computing systems (mostly UNIVAC, IBM and DEC machines) and exchange (compatibly with portability issues, at that time still far from being solved) the software systems that were available in the various sites: LISP interpreters, algebraic manipulation systems such as Macsyma and REDUCE, reasoning systems such as PLANNER, question answering systems, automatic theorem proving systems, etc.

In those meetings we felt the need to consolidate our domain of research; we realized that computer science was progressing on the two legs of technology and applications but it was clear to us that behind machines and programs the concepts of computing had still to be fully understood and that progress in the theoretical foundations was needed in order to improve efficiency and correctness of systems and applications.

At the current level of knowledge it was a miracle that such complex applications as those governing the moon landing missions or nuclear plant control systems were actually working without accidents. In 1968 the concern regarding efficiency and correctness of computer systems reached alarm level. Fritz Bauer (Fig. 4.8), one of the pioneers of European informatics[15], was then asked to organize an international conference under NATO spon-

---

[15] Friedrich Ludwig (Fritz) Bauer (1924–2015) is known as one of the fathers of German informatics. Originally a theoretical physicist he started to teach mathematics and then computer science. He is credited with having invented the stack method for evaluating

Fig. 4.8 Friedrich Ludwig (Fritz) Bauer

sorship. "After meticulous preparation of the agenda, a conference of 60 participants was subsequently held from 7th to 11th October 1968 in Garmisch. The candid conference report did not fail to have a salutary effect on the development of informatics in the NATO countries and beyond. Nowadays it is taken for granted in the software industry that whatever tools are available should be used properly (with *ingenium*). *The computer, one of the greatest inventions of engineers, has to go the complete way of engineering to its end*".[16]

Let us see how Brian Randell, who organized the conference together with Peter Naur and Fritz Bauer, describes the situation in 1968. "The idea for the first NATO Software Engineering Conference, and in particular that of adopting the then practically unknown term 'software engineering' as its (deliberately provocative) title, I believe came originally from professor Fritz Bauer. Similarly, if my memory serves me correctly, it was he who stressed the importance of providing a report on the conference, and who persuaded Peter Naur and me to be the editors. As I and other participants have since testified, a tremendously excited and enthusiastic atmosphere developed at the conference. This was, as participants came to realize, the degree of common concern about what some were even willing to term the "software crisis", and general agreement arose about the importance of trying to convince not just other colleagues, but also policy makers at all levels, of the seriousness of the problems that were being discussed."

The conference had indeed a strong impact and was soon followed by a second conference in Rome, organized in 1969 by Brian Randell and J. N. Buxton. The development of software engineering methodologies and support environments seemed to be the key to improve the software production process and to achieve a better reliability of systems. But, in the view of computer science theoreticians, as any other 'engineering domain' (like me-

---

arithmetical expressions. He took part in the definition of Algol 60 and in the organization of the first events devoted to software engineering in Europe.

[16] Friedrich L. Bauer, *Origins and Foundations of Computing*, Springer, 2010

chanical or electrical engineering) no valid practical developments could be
expected without the parallel development of a scientific background leading
to a better understanding of the basic concepts of computing. This is what
the new science was to be aimed at.

In particular the new research domain of semantics of programs was expected to have a dramatic influence on the quality of software production.
In fact, in order to assess the correctness of the behaviour of a computer
program, one of the first issues that had to be addressed was: what is the
meaning of a program? Indeed, in order to understand what a program meant
at that time it was necessary to make reference to a specific implementation
of the programming language. Not rarely, two different implementations of
the same program might have two different meanings and might produce
two different behaviours. In their fundamental paper *Toward a Mathematical
Semantics for Computer Languages*, which appeared in the *Proceedings of
the Symposium on Computers and Automata* (organized in April 1971 at the
Polytechnic Institute of Brooklyn), Dana Scott and Christopher Strachey[17]
write: "Compilers for high-level languages are generally constructed to give
the complete translation of the programs into machine language. As machines merely juggle bit patterns, the concepts of the original language may
be lost or at least obscured during this passage. The purpose of a mathematical semantics is to give a correct and meaningful correspondence between
programs and mathematical entities in a way that is entirely independent
of an implementation". A similar observation appears in the paper *Denotational Semantics* by Peter Mosses: "Currently, most programming language
standard documents attempt to define semantics by means of informal explanations. This is in contrast to syntax, where formal grammars are routinely
used in standards (in preference to informal explanations). However experience has shown that informal explanations of semantics, even when they are
carefully worded, are usually incomplete or inconsistent (or both) and open
to misinterpretations by implementors. They are also an inadequate basis for
reasoning about program correctness and totally unsuitable for generation of
implementations".

In the 1960s and the early 1970s various approaches to semantics had been
introduced. On one side, following the so-called axiomatic approach of Floyd
and Hoare, as we have seen, it was proposed that the behaviour of a program
could be defined in logical terms by specifying the conditions that should be
satisfied after the execution of a portion of a program in the hypothesis that
suitable conditions were satisfied before. Other approaches, e.g. McCarthy's
approach, were instead based on the definition of an abstract machine that
was supposed to interpret the program. In the mathematical approach, in order to circumvent the limitations of other approaches, the effort was to take
a more abstract point of view and to associate a mathematical object (called

---

[17] Christopher Strachey died in 1975 at the age of 59. He had a very important role in the
history of computer science for the work he did on the very early computers (like the ACE
and the Ferranti Mark1) and for having initiated the field of program semantics.

'a denotation') to each phrase of the language; in such a way a mathematical meaning (a function) consisting of the composition of the more elementary denotations can be associated to complex portions of a program. After the first paper published by Christopher Strachey in 1964 in the Proceedings of the Working Conference on *Formal Language Description Languages for Computer Programming*, between 1969 and 1971 a series of papers by Dana Scott and by Scott and Strachey laid the foundations of mathematical semantics (later known as denotational semantics). An important role in this context was played by Scott's two papers *Outline of a Mathematical Theory of Computation* (which, as we already said, appeared at the 4th *Annual Princeton Conference on Information Sciences and Systems* in 1970), and *The lattice of Flow Diagrams* (presented in the 'virtual' symposium *On Semantics of Algorithmic Languages* whose proceedings were edited by Erwin Engeler in 1971[18]). In these papers lattices and continuous functions over lattices are proposed as the suitable mathematical domains to adopt as a basis for specifying the semantics of programs, the fundamental breakthrough that allowed us to reach a precise definition of the semantics of recursive programs as fix points of functional equations over such domains. This work, and his previous contribution on finite state automata, jointly realized with Michael Rabin, earned Dana Scott (Fig. 4.9) the Turing Award in 1976.

Fig. 4.9 Dana Scott

---

[18] Erwin Engeler, born in 1930, is a great Swiss logician and computer scientist. In the 1950s he had been one of the students of Paul Bernays at ETH in Zurich, then for several years he was a professor in various US universities. From 1972 until retirement he was professor at ETH. The proceedings of the virtual symposium contain contributions by most of the active researchers in semantics and logics of programs: among them de Bakker, Hoare, Knuth, Manna and Scott. In the preface Engeler says: "During the last few years, a number of interesting results and promising ideas have been generated in the area of semantical aspects of programming languages. We felt that we would do a real service to this emerging field by calling a write-in symposium and collecting a number of representative contributions covering the various aspects".

Despite the fact that the theory of computing was so young at the beginning of the 1970s, the field had already incredibly expanded. Today, fifty years later, the two communities, the computer scientists who work on the design and analysis of algorithms and on computational complexity and those working in logics and semantics (often identified as the Track A and Track B communities, from the subdivision of topics frequently applied in journals and conferences) are separated by different mathematical backgrounds, different techniques, and different languages. Still, looking back now at what theoretical computer science meant for us in those days, it appears that we used to see the field as a unique one in which all research directions, all approaches used to create formal models of aspects of computing (languages, computation models, computational complexity, semantics and correctness of programs etc.) were merging in a mathematical cultural universe whose bases rested on algebra, logic and discrete mathematics.

In Spring 1971 Ugo Montanari and Giorgio Levi, who used to come frequently to Rome to collaborate with CNR Laboratories on medical image recognition problems, told me that they wanted to talk to me. I had met Ugo in Stanford and that had been the beginning of a deep friendship, so I was very excited when he told me that in agreement with Antonio Grasselli and his other colleagues in Pisa he wanted to invite me to join the crew that was starting off the new degree in computer science. I should transfer my CNR position from IAC in Rome to IEI in Pisa; in particular, I was proposed to teach a course in computability. The offer was overwhelming and I spent hours and days in discussion with my family to decide what to do.

Finally I took the train to Pisa to present my CV and my application for teaching the computability course to the faculty, but when I met Fabrizio Luccio[19] at the Istituto di Scienze dell'Informazione I was still undecided. Fabrizio, who also, since then, is in the short list of my best friends, at that time had some official role in the faculty. Born in Lybia, he lived in Rome during his childhood and was the brother of one of my schoolmates at primary school. That day, in his office in Corso d'Italia, we had a long discussion on the perspectives of theoretical computer science in Italy and he tried his best to convince me to apply for the position in Pisa but at the end the idea not to leave Rome and not to leave my mother who had become a widow only a few months before, prevailed. I withdrew the application that was already in his hands.

I still don't know whether it was good or not for me personally to remain in Rome, but this decision forced me to think that something should be done to promote theoretical computer science studies in Italy. This idea became

---

[19] Fabrizio Luccio, born in 1938, is one of the most productive leaders of Italian computer science. His scientific interest is primarily in the field of algorithms and data structures. Professor in Pisa since 1971 he has also been often invited as a professor in the US and in several other countries. Among his students are excellent scientists such as Paolo Ferragina, Roberto Grossi, Linda Pagli and Geppino Pucci.

stronger, as we will see, after the important meeting I had in Israel that summer.

In July another of the summer schools that contributed to the making of the new science and led to the creation of the European computer science community took place in Marktoberdorf, in Bavaria. I had the chance to be there.

The school was organized by Fritz Bauer; again the list of lecturers was impressive. Among them there were five Turing Award winners: Tony Hoare, who lectured on elements of programming languages, Niklaus Wirth, who spoke about implementation of Pascal, Edsger W. Dijkstra, who presented a course on parallel programming, and finally Alan Perlis and Ole-Johan Dahl. Besides, in a course devoted to operating systems, Per Brinch Hansen presented the material that a few years later he put in his book, one of the most successful on this subject and, in a course devoted to information retrieval, Rudolf Bayer lectured about B-trees, the powerful data structure that, as we already mentioned, he had just invented jointly with Ed McCreight. Finally, John Reynolds gave a thorough presentation of Dana Scott's approach to the mathematical semantics of programs.

According to the notes that Dijkstra later posted in his blog, the school was attended by 80 students from 15 countries. Dijkstra says "Some participants were very theoretically inclined, others more practically minded". Also, from Dijkstra's perspective, the summer school presentations were about "consolidation" (i.e. confluence of various technical developments into a unified stream toward structured programming).

To me, the most interesting lectures presented at the school appeared to be those given by John Reynolds[20] (Fig. 4.10). It may be that I was influenced by my old studies on λ-calculus but the elegant and powerful construction leading to the precise definition of the semantics of recursive programs as fix points of continuous transformations over semi-lattices impressed me deeply. Undoubtedly Scott's definition of mathematical semantics and the series of results that followed this approach were the most important theoretical achievements marking the early 1970s.

John Reynolds' lectures were crisply clear and I started to consider (somewhat naively, I must admit) whether this framework be suitable to establish a correlation between the semantics and the complexity of programs through an approximation notion over domains.

After the school in Marktoberdorf the next big event that summer was a conference that took place in Haifa in the subsequent month of August. The conference was organized by Zvi Kohavi and Azaria Paz, both from the

---

[20] John Charles Reynolds (1935–2013) was one of the most important American computer scientists involved in the field of program semantics. From 1970 to 1986 he was a professor at Syracuse University. Subsequently he became professor at Carnegie Mellon University. His research interest were always in the field of programming languages and formal semantics (in particular λ-calculus based semantics). Among his PhD students of note is Benjamin Pierce.

Fig. 4.10 John Reynolds

Technion, and its title was *Theory of Machines and Computations*. The list of topics covered at the conference (which besides computability, automata and formal languages, included fault detection, combinational circuits, and sequential circuits) shows that we still were in the transition between the old time in which the most relevant theoretical issue in computer science was switching theory and the subsequent period in which the full spectrum of topics that later was called theoretical computer science was displayed. At Haifa I had the pleasure to meet again my mentor Manuel Blum and several colleagues I had met in the US: Albert Meyer and Ed McCreight (who presented their work on computationally hard pseudo-random zero-one valued functions), Allan Borodin (who presented a paper on the optimality of Horner's rule for computing polynomial forms), Dennis Tsichritzis (who at the time was working on the properties of various kinds of subrecursive programming languages)[21]. The conference gave me also the opportunity to meet other scientists engaged in complexity research, first of all Azaria Paz[22], our Israeli host, a very good friend of Corrado, at that time working on probabilistic automata, and John Hopcroft, one of the leading figures at Cornell, whom I had not met during my visit to that university the year before because at that time he was visiting Stanford University where he was producing his remarkable joint work with Robert Tarjan. At the conference John Hopcroft presented one of his landmark papers: *An n log n Algorithm for Minimizing States in a Finite Automaton*, a result achieved during his stay at Stanford; with its 900 citations this is certainly the most cited paper

---

[21] In the 1980s Tsichritzis became more famous for his work (and his textbooks) on operating systems and on databases.

[22] Azaria Paz is one of the greatest Israeli computer scientists. He took his PhD at Hebrew University under the advisorship of Michael Rabin. From the early 1970s he was a professor at the Technion. His research interests span from automata theory to graph and algebraic algorithms and also to aspects of automatic reasoning. In 1971 he had just published his most famous book *Introduction to Probabilistic Automata*.

among those presented at the conference especially because, as far as I know, it was never published in a journal.

That was my first visit to Israel and I was overwhelmed by the history and the society of that country. A spectacular night concert on the beach in the archeological site of Caesarea was one of the key social events beside a one-day excursion in Galilee, around the Sea, visiting 'kibbutzim' and talking to 'kibbutzniks' who still had memories of World War II. The excursion stretched up to the Golan Heights that had been recently occupied by Israel in the Six Days War, adding to the journey the feeling of witnessing important moments in Israel's life.

The conference was also attended by Maurice Nivat[23] (Fig. 4.11). In the subsequent years Maurice Nivat would become one of the most influential European computer scientists. In 1965 his encounter with Schützenberger, to whom he would remain strongly attached, had marked his life and introduced him to the elegance of algebraic approaches to the study of formal languages. In 1967 he had defended his 'Thèse d'Etat' on Chomsky languages and soon after he had become a professor at Paris University (at Paris VII after the 1970 split of the Paris universities). In 1970 Schützenberger introduced him to IRIA where in 1971 he had created the project: *Formal Semantics of Programming Languages*[24].

When I saw him in Haifa he had long hair and was wearing bell-bottom pants with white and blue stripes. On one of the last days he approached me. He wanted to tell me that the next winter he would be in Rome to give some lectures upon invitation from the 'Istituto di Alta Matematica'. He

Fig. 4.11 Maurice Nivat (right) and Maurice Gross at the first ICALP (1972)

---

[23] Maurice Nivat was born in 1937. His death in September 2017 was a terrible loss for the entire theoretical computer science community. His role in the field has been extraordinary, both, as we will see in these notes, for the many initiatives that he organized to promote the foundational and scientific side of computer science and for his scientific activity. Many outstanding French computer scientists were his students; among them Luc Boasson, Bruno Courcelle, Pierre-Louis Curien, Philippe Flajolet, Gérard Huet.

[24] A vivid account of Maurice Nivat's scientific life was written by P.-L. Curien: *Une brève biographie scientifique de Maurice Nivat*, Theor. Comp. Sc. 281, 2002.

also told me that he was aiming at creating some kind of European organization in order to promote research on theoretical issues in computer science. Recently something similar had been created in the field of biology, the European Molecular Biology Organization (EMBO) and he wanted to propose to Brussels that also the development of theoretical computer science could be accompanied by a European initiative. He announced that I would be invited to take part in a meeting to be held in Brussels some time in winter 1971–1972.

Such unexpected news excited me. It was coming at the right time and it was incredibly in sync with the activity that we were planning in Italy to support the new science and with the events we had to face in the coming fall. The Italian scenario concerning the theoretical foundations of computer science has been described already. Several excellent research groups but no unified organization, no common ground for pushing the field forward, no regular event or occasion to present and compare each other's work. More generally the scenario of the whole area of computer science was much alike. The main university research groups were still under the suffocating embrace of other academically stronger groups: electrical engineers in Faculties of Engineering and mathematicians or physicists in Faculties of Sciences. Despite good personal relations, our colleagues who formed the first generation of Italian computer scientists did not have many occasions to engage in a synergic effort for consolidating the field. Of course most of the Italian computer scientists were members of AICA, the Italian informatics association, that used to organize an annual conference (although with modest scientific relevance) and many of them were members of the ACM (the American *Association for Computing Machinery*) but there were no organizations that could help the growth of the identity of the new science and could give visibility to computer science in the academic and national research environments.

In universities the creation of the Laurea in Scienze dell'Informazione had been an important acquisition but it was limited to the faculties of sciences of four universities. In the faculties of engineering computer science courses were only delivered for special curricula within the framework of the degree in electrical engineering or in electronics. In 1971, at Rome University, in the curriculum for the degree in electronics only one course out of 27 was a computer science course. This course ('Electronic Computers') was taught by Paolo Ercoli who at that time was not even a faculty member but still a CNR researcher.

The only concrete sign of interest toward computer science had been expressed by Antonio Ruberti (Fig. 4.12), the scientific leader of the group of control theory and automation, who had just started a graduate program in Engineering of Automatic Control and Computing Systems, almost at the same time at which he was creating the CNR center with the same name that was mentioned before. Ruberti had invited a few young computer scien-

Fig. 4.12 Antonio Ruberti

tists to join, among them Daniel Bovet, Giacomo Cioffi and me[25], who shyly tried to define a computer science curriculum in the program. In particular Giacomo and I shared the lectures of the course on Combinatorial and Sequential Systems, essentially a course devoted to automata and switching circuit theory in which I tried to include basic notions of formal languages and computability.

At CNR, the Italian National Research Council, as we pointed out before, computer science was slowly developing but still under the umbrella of other disciplines: mathematics, physics or engineering. Although first class studies in computer science were carried on in various CNR institutes and research centers, no official formal body was designated with the role of funding research in computer science. Despite the existence of a Commission for Informatics and of a so-called Special Project for Informatics all decisions were taken by the members of the consulting committees of the well established 'major' disciplines and depended on the interpretation that the members of such committees were giving to informatics, and in most cases informatics was perceived as instrumental for applications in other fields.

We were very unhappy about this situation and we were looking with interest to the signs coming from other European countries that appeared ready to develop informatics. In particular France seemed to proceed with greater decision in this direction. In 1966, following the takeover of Bull by General Electric, the French government had launched a strategic plan, called *Plan Calcul* for the development of the computer industry in France and for a big effort in research and education. In this framework, in 1966 a new governmental body responsible for informatics had been established, the *Délégation à l'Informatique*, directly reporting to the Prime Minister, and a new computer company (the *Compagnie Internationale pour l'Informatique*,

---

[25] The offer that Ruberti made to me to teach in his *School on Automatic Control and Computing Systems* was just the first step of a long productive journey that I traveled close to him, for which I will always be indebted.

CII) was launched. Besides, for promoting research in informatics the *Institut de Recherche en Informatique et Automatique* (IRIA) had been created. Under its first President Michel Laudet (in charge from 1967 to 1972) IRIA immediately gained a leading role in computer science research in Europe and became a model at which computer scientists in other countries were looking for promoting their science.

In the Fall of 1971 an important event upset the Italian computer science scene. At the AICA annual congress that was taking place in Genoa on October 21–23, Antonio Ruberti launched what we, computer scientists, saw as a 'takeover bid' for Italian informatics. He proposed the creation of a joint formal organization that could put together researchers in automation, bioengineering and informatics. Named GRABI (*Gruppo di Ricerca in Automatica, Bioingegneria ed Informatica*) the group would have been responsible for promoting these research fields with respect to CNR, which was the main (at that time the only!) funding agency for universities. This move would have helped Italian computer scientists to escape from the deadly embrace with other powerful disciplines but it soon appeared that as a consequence we would be subject to the cultural and scientific design of our colleagues in control and automation who, although aware of the big relevance of informatics, viewed our discipline just as a part of the largest and all-encompassing field of system science.

A small digression regarding system science is here necessary, since the approach inspired by this discipline had a big influence on the scientific, social and economic culture in the 1970s, both in the Western world and in the East European block. Systems science is an interdisciplinary research field, whose first steps can be traced in the 1950s, after WWII, and whose motivation was based on the need to resort to mathematics and formal approaches for analyzing, understanding and regulating the behaviour of complex systems that are present in nature, in society and in economics, as well as in science and engineering.

Among the first scientists who promoted the system science paradigm we can remember Ludwig von Bertalanffy, Margaret Mead, Anatol Rapoport and many others. A typical example of the system approach to improve the behaviour of social systems can be seen in the works of Jay Wright Forrester (born in 1918) who in the 1970s became famous with his book 'Urban Dynamics' which inspired a political and scientific debate on the feasibility of modeling social systems and introducing mathematical regulatory methods for solving social problems.

Antonio Ruberti, who in 1964 had become full professor of Control Theory at the University of Rome (the first Italian full professor in this new scientific field) was always fascinated by the interdisciplinary power of the mathematical models and methods of control theory and for a long period in his life was a strong advocate of the system science paradigm. Still in 1982, when the limits of the approach when applied to social and economic systems were becoming evident, he was saying: "System science helps to isolate a phenomenon from the universe it belongs to and relates it to the same universe by means of entities that can be modified (input) and entities that can be observed (output): the relation between such entities is the system associated to the phenomenon. The system therefore appears as a general concept of mathematical model and has in itself the potential to be employed in the most diverse sectors. Its generality releases the definition from specific disciplinary contexts and allows that it be widely adopted"[26].

After becoming professor of Control Theory, Ruberti started to work toward building a university department for his field, control and automation (internationally known also with the name of 'automatica'). In 1969 he finally succeeded to create the *Istituto di Automatica*, separated from the electrical engineering institute, and in the same year, upon his proposal, CNR created the research center for *Automatic Control and Computing Systems* (CSS-CCA, Centro di studi sui sistemi di controllo e calcolo automatici). Both institutions were indeed covering a large scientific domain and their scope included the field of informatics as well as the field of control and automation. Rome was indeed becoming the capital of control and automation in Italy but in Ruberti's vision his project would not have been complete until the union between automation and computer science was realized at the national level with the creation of a formal organization (such as, for example, a CNR research group) encompassing both fields.

Ruberti's proposal was not favorably received by Italian computer scientists, although some colleagues belonging to faculties of engineering were more ready to accept the offer. In particular those among us who were more theoretically inclined felt that the umbrella that Ruberti was offering would have left our field in a subordinate position with respect to the emerging domain of system science. Besides, for us the idea that informatics was just a subfield of system science was also culturally and scientifically wrong. Clearly, the modeling power of informatics could reach far beyond the mathematical modeling capability offered by the systems of differential equations used by system and control theorists. In any case, also taking into consideration the successful initiatives created by Ruberti in Rome, a closer look at the research programs of the two institutions (the Istituto di Automatica and the CNR research center CSSCCA) allowed us to put in evidence that, in his vision, informatics was just one of the many subfields of system science[27].

---

[26] See *Antonio Ruberti*, by Claudio Gori Giorgi, Sapienza Università Editrice, 2014.

[27] Still in 1974, when Ruberti invited me to join CSCCA to develop the Informatics Section of the Center, the research activity was organized in five sections and only one of them was devoted to informatics while four were devoted to various aspects of systems science: system theory, control theory, servomechanisms (sensing and actuating devices), and bioengineering.

Chapter 5
Informatique théorique

Ruberti's proposal had in any case a positive effect. Not only it had started a major debate that involved almost all Italian computer scientists but it also contributed to determine in this community a sense of identity that had been lacking until that moment. The first reaction came from theoretical computer scientists.

Already in the same month, October 1971, a few days before the AICA conference where Ruberti presented his project, on the occasion of a workshop on 'Automatic Information Processing' held in Florence from the 11th to the 13th of October, a meeting had been called with the aim to create a group of researchers of theoretical computer science (with the name GRIT: Gruppo Ricercatori di Informatica Teorica). Researchers from Florence, Pisa, Rome, Bologna and Naples, altogether 23 people, had taken part in the meeting. Only Milan and Turin were absent, for various reasons. An important role in the meeting was played by Alfonso Caracciolo. At that time he was a member of the CNR Informatics Commission, but he also had responsibilities in Brussels and for this reason he was aware (indeed he was one of the promoters) of the efforts that were starting at European level for creating a European initiative in the field of theoretical computer science, the initiative that had been announced to me by Maurice Nivat in Israel, a few months earlier.

Let's see Maurice Nivat's account of how Caracciolo and himself decided to move toward the creation of a European association: "It begins in 1968, the year of the big turmoil in French universities, I had got my doctoral degree in 1967 with a dissertation on Chomsky languages prepared under the supervision of Marcel-Paul Schützenberger (Fig. 5.1) who had been appointed as one of the four directors of the institute IRIA which was created around that time by the General De Gaulle, the head of the French State angry with the embargo put on the most advanced computers by the United States of America. I was sitting as a member on a panel of a dozen people gathered by OECD to think about the future of the new discipline called in English *Computer Science* and in French *Informatique* which was developing fast

© Springer International Publishing AG, part of Springer Nature 2018
G. Ausiello, *The Making of a New Science*,
https://doi.org/10.1007/978-3-319-62680-2_5

Fig. 5.1 Marcel-Paul (Marco) Schützenberger

though only a few people really knew what was the thing about. In this panel I met an Italian fellow, Alfonso Caracciolo di Forino, who impressed me a lot for many reasons: he belonged to the noble Neapolitan family of the admiral Caracciolo who was hanged in 1799, in spite of all laws of war, at the instigation of Nelson ... he was a very charming and talkative man and he was active inside the European Community (EEC) trying to have a European software institute created with the major computer manufacturers of that time in Europe, the Italian Olivetti, the British ICL, the German Siemens, and the French Bull company. During our meeting in the Château de la Muette, in Paris, he told us of the difficulty of the negotiations between these companies, which eventually never ended in whatever. And that is when chatting after a meeting he told Schützenberger and myself that, maybe, it would be possible to create a European entity dealing with the theoretical aspects of computer science since theory can be pursued with little money and does not raise the unsolvable financial and economic problems which were at stake when talking about European cooperation in the software industry. He suggested that we create a European association.[1]"

At the Florence meeting Caracciolo gave us some more details on the European plans regarding the creation of a European association and in particular explained that it would be very important to create a corresponding organization on the Italian side. So the main issue at that point concerned the framework in which the group should be created. The easiest step would have been to form a kind of 'interest group' within the Italian computer association, AICA, but indeed the real goal was to take an initiative that could enhance the visibility of theoretical computer science with respect to the funding institution, CNR. At the end of the meeting we decided to go

---

[1] Maurice Nivat, *The True Story of TCS*. Theoretical Computer Science, Special Issue for TCS 40th Anniversary.

on in the direction of the creation of an AICA interest group for 'theoretical computer science' but we also decided that we would (i) take steps to get all major Italian groups working in theoretical computer science involved, in view of organizing schools and symposia and of coordinating the invitation of foreign scientists, (ii) establish contacts with CNR for preparing the ground for a more formal initiative that could be recognized by the Consulting Committees of Mathematics and Engineering and that could play a role in the project to be carried on at the European level.

So when Ruberti's offer to gather all researchers in system theory and in computer science in a unique group was presented in Genoa, theoretical computer scientists had already taken an alternative direction and this move had a strong impact in changing the attitude of most Italian computer scientists toward Ruberti's proposal. On December 6 a second meeting took place at IAC in Rome. Twelve people were present, representing all major Italian groups: Corrado Böhm from Turin, Gianni Degli Antoni and Marco Somalvico from Milan, Alfonso Caracciolo, Luigia Carlucci Aiello and Ugo Montanari from Pisa, Giorgio Germano, Andrea Maggiolo and Francesco Lauria from Naples, Giorgio Ausiello, Alfonso Miola and Marisa Venturini Zilli from Rome. I was asked to coordinate the GRIT initiative and, in this role, to inform the Commission for Informatics of CNR of the creation of the AICA interest group (also in view of asking the Commission to provide financial support for schools and conferences to be organized in the future), and, at the same time, to send an invitation letter to all colleagues who might be interested in joining the group.

But, in the meantime, at European level things were moving much faster than we could imagine. In mid-December I received a letter from Mr. Desfosses, secretary of the 'Special Group for Computer Science Education' that had been established by the Commission of European Communities (CEC) inside the Group PREST (*Politique de Recherche Scientific et Téchnologique*). I was invited to take part in a meeting to be held in Brussels on January 27 and 28, 1972. The meeting, whose official title was 'Cooperation in the field of theoretical data processing', was called by the Special Group with the aim to "define actions for the promotion of the field of theoretical computer science in support of computer science education and research ... and to develop cooperation among European Universities". The meeting was chaired by Alfonso Caracciolo di Forino. The letter also contained two enclosures, both prepared a few months before by Maurice Nivat and his colleagues in Paris VII[2]: the proposal for an initiative leading to an 'international multi-

---

[2] The role of Maurice Nivat in the birth and growth of the new science is well described in a French blog that starts with the following words: "En 1970, Maurice Nivat, jeune mathématicien, et Marcel-Paul Schützenberger, son père spirituel, nommé depuis peu au directoire de l'IRIA (ancêtre de l'INRIA) avec Henri Boucher, André Lichnerowicz et Jacques-Louis Lions, préparent un événement pour le moins inédit: une conférence de presse pour annoncer la naissance officielle de ce qu'ils baptisent 'l'informatique théorique'!" "Je ne sais pas ce qui nous a pris ce jour-là" reconnaît Maurice Nivat. "Toujours est-il que prés de 40

university' and another 12-page document, dated July 22, 1971 whose title was: 'Rapport preliminaire sur l'informatique théorique'.

Both documents are very important for our historical survey on the development of the theory of computing. The first document prepared by Maurice Nivat was inspired by another European initiative that had appeared very successful and a possible model to follow: the European Molecular Biology Organization (EMBO) created in 1966[3] with the aim to promote the development of this discipline and the international scientific cooperation at European level among research institutions and universities. This document can be considered by all means the first draft idea of the creation of the European Association for Theoretical Computer Science. The proposal, for which we can give credit to Maurice Nivat, Louis Nolin and Marcel-Paul Schützenberger, is well illustrated in a letter of Professor François Bruhat, dated June 24, 1971, in which the Director of the UER Mathématique of Paris VII explains (presumably to Alfonso Caracciolo), the motivation for a cooperation among European universities "dans le domain de l'enseignement de l'Informatique et dans celui de la recherche en Informatique Théorique" (see Appendix C). From the point of view of scientific research the objective of the cooperation would be to promote the exchange of information and of scientific results in the field and to organize visits of young researchers in laboratories of other countries (a kind of ante litteram Marie Curie fellowship program) and specialized schools and conferences. Besides Paris VII, the sites indicated as possible partners in the initiative were: Saarbrücken and Munich in Germany, Rome, Pisa and Turin in Italy, Amsterdam and Brussels for The Netherlands and Belgium, Paris VIII and Toulouse for France. Furthermore, in the United Kingdom (at the time not yet a member of the EEC) the universities of Warwick, Edinburgh and Colchester would be invited to join the project.

The content of the second document, also prepared by Maurice Nivat and his colleagues, is of remarkable interest because it is one of the first attempts to define the field of theoretical computer science and to analyze its relationship with the more practical aspects of computing systems and their applications (see Appendix B).

The document (a preliminary version of which had already been sent to the French Delegation for Informatics) starts with the words: "Dans les pages qui suivent nous tentons d'isoler et de délimiter aussi précisément que possible un domaine de recherche que, faute de mieux, nous appelerons Informatique théorique". In a sense, although, as we have seen, since the mid-1960s various conferences and summer schools had already been devoted to this domain and in the US university curricula in computer science were already strongly

---

ans plus tard, la discipline existe bel et bien et que le nombre de chercheurs concerné a été multiplié par cinquante si ce n'est cent." (https://interstices.info/jcms/c_32847/maurice-nivat-une-vision-a-long-terme-de-la-recherche-en-informatique)

[3] It is just a nice coincidence that also the roots of the EMBO initiative can be traced back to a successful conference held in Ravello in 1963.

inspired by the theory of computing concepts, this was the first time in which the 'new science' was characterized in an official document and its major subdomains defined[4].

First of all the document makes a disclaimer: it specifies what is not 'informatique théorique'; a variety of disciplines that had developed in relation with computing systems and that can be considered in their own right part of informatics (numerical analysis, operations research, control theory, artificial intelligence, ...) in fact are not taken into consideration since they are more interested in exploiting the power of computers rather than investigating the properties of computation processes. Then the document continues: "Au contraire, c'est cette notion de calcul, dans ses aspects les plus variés et qui pour una large part échappe à la mathèmatique classique qui est au centre de ce que nous appelons Informatique théorique". After putting in evidence that the need for a better understanding of the notion of computation arises from various applied domains of informatics with the aim to improve correctness and efficiency of computation processes (notably such need had been clearly expressed in the conference on 'Software Engineering' that had been held, as we saw, in 1968) the document says that "... cette discipline, croyons-nous, conditionnera dans une large mesure le développment de toute l'Informatique à long terme ...".

Three subdomains of theoretical computer science are presented in the document, each one illustrated in detail and with examples taken from the first successful achievements of those early days:

- Algorithmics, in particular with reference to non-standard computation models (i.e. parallel computation models, multilevel storage structures, etc.), to algorithmic techniques (such as 'divide and conquer' algorithms for integer and matrix multiplication), to analysis of running time of algorithms (such as Knuth's approach to computing the execution cost of an algorithm in the average case) and to complexity of optimization problems (such as the traveling salesman problem).
- Automata theory and formal languages, moving on one side from the first results of Kleene and Rabin and Scott and on the other side from the work of Chomsky and Schützenberger and having as central chapters the theory of codes, the theory of finite automata and of regular languages (in the French terminology, 'languages rationnels') which also included Kron and Rhodes' algebraic approach, the theory of pushdown automata and

---

[4] In the paper *ICALP, EATCS and Maurice Nivat* (Theor. Comp. Sc., 281, 2002) Grzegorz Rozenberg and Arto Salomaa say: "In our days it is hard to visualize any more the scientific landscape and academic surroundings in the early 1970s in fields now referred to as computer science. Everything was rather unorganized and scattered. Automata theory was already very advanced. On the other hand, what was labeled as theoretical computer science or theoretical informatics varied enormously and could be, for instance, numerical analysis or some parts of theoretical physics. Maurice Nivat apparently wanted to have some order and uniformity, and was also very successful in his efforts to achieve this goal."

of context-free languages (again in French terms, 'languages algébriques') and the theory of 'tree automata' ('automates d'arbre').

• Formal semantics of programming languages, where the document, while underlining the difficulty of assigning a precise mathematical meaning to a program (e.g. by determining the abstract function that it computes), illustrates the first attempts that had been carried out until that moment, such as Floyd's logical approach based on pre- and post-conditions, McCarthy's approach for functional languages, the approaches based on λ-calculus (Dana Scott) and combinatory logic (Nolin) and the syntactic approach based on program schemata and on program transformations[5].

Finally, the document makes an interesting remark on the importance to study semantics and correctness of operating systems (in line with Dijkstra, Wirth and Naur's works), the real benchmark for the theory of computing: "... l'étude des systèmes doit devenir, s'il n'est déja, le lieu privilégié d'application de l'Informatique Théorique".

Nivat's documents arrived in Italy at the most appropriate moment, when we were trying to define the role and the scope of GRIT and we wanted to establish actions for the promotion of theoretical computer science. On December 28 I sent an invitation letter to all colleagues who might be interested to join GRIT, calling for another meeting to be held at Milan Polytechnic on January 24 in view of the international meeting that would be held in Brussels on January 27 and 28, and I sent to this larger circle of people copy of Nivat's document on 'informatique théorique'.

The year 1972 is a milestone in the history of theoretical computer science both for the events that took place in that year and that, as we are going to see, marked the beginning of a new phase in the development of the new science, and for the scientific results achieved in that year in Europe and in the States, results that made the history of computer science and are still cited as pillars in their own respective fields[6].

---

[5] "Maurice's interest in semantics goes back to a meeting of the IFIP WG 2.2 in Colchester in 1969 on the semantics of programming languages – a subject that was just starting at that time. Maurice learned from David Park and Mike Paterson about program schemes. This led to a large body of work for Maurice and a number of his students." (Curien, P.-L. *Une bréve biographie scientifique de Maurice Nivat*, Theor. Comp. Sc. 281, 2002). Nivat's algebraic approach to semantics was essentially a sort of purely syntactic operational semantics in which the basic domain consisted of the Herbrand universe, the 'magma libre' as he used to call it.

[6] It is interesting to observe that although the state of computer science in 1972 is well represented by a volume of the *Communications of the ACM* celebrating the 25th year of the association, the special issue still misses to catch the most important research directions in the theory of computing. In the section 'Fundamentals' of the special issue five domains are identified as the key themes to address in future years but only two of them correspond to what we would now call theoretical computer science: Michael Arbib's paper concerning applications of automata to the theory of the brain, and Zohar Manna and Jean Vuillemin's paper devoted to the fixpoint approach to the theory of computation. No mention of advances in algorithm design, no mention of computational complexity.

We have already mentioned that one of these achievements concerns the invention of B-trees by Rudolf Bayer and Ed McCreight (the paper *Symmetric Binary B-trees: Data Structures and Maintenance Algorithms*, was published in *Acta Informatica* in 1972), a result that had a strong impact on the design of the physical organization of databases. In the same year two events made the progress in the field of programming languages impressive. The publication of the book *Structured Programming* by Dahl, Dijkstra and Hoare had a fundamental role to transform this programming discipline from what had previously been just a set of theoretical principles into a standard program design methodology. Also from 1972 is the first implementation of PROLOG (from the French words PROgrammation LOGique) by Alain Colmerauer and Philippe Roussel[7], a language that introduced the declarative style of programming and that had a crucial role in establishing a bridge between artificial intelligence and databases in the subsequent decades.

In 1972, the proceedings of the *Courant Computer Symposium on Data Base Systems* (Randall J. Rustin Ed.) were also published. The symposium had taken place the year before, in May 1971, and was one of the first events in which theoretical issues regarding databases were addressed, prompted by the recent work that had been done by Edgar Codd[8] (Fig. 5.2) at IBM with the introduction of the relational model of data. In particular at the symposium Edgar Codd presented the papers *Further Normalization of the Data Base Relational Model* and *Relational Completeness of Database Sublanguages*.

Also in the field of semantics important events took place in 1972. First the Courant Institute of NYU organized a *Symposium on Formal Semantics* (again, probably, the first big event devoted to this new research domain) that gave rise to the book *Formal Semantics of Programming Languages* (Randall J. Rustin Ed.) including contributions of Dana Scott and of other major scientists involved in the field. Also the *ACM Conference on Proving Assertions about Programs* that took place in the same year placed a big emphasis on the issue of program semantics and again the relevance of Scott's approach was acknowledged in Milner's paper *Implementation and Application of Scott's Logic for Computable Functions* in which he presents LCF, the logical system for the proof of program properties based on Scott's ideas orig-

---

[7] Robert A. Kowalski, *The early years of logic programming*. Communications of the ACM 31:38 (1988).

[8] In 1970 Edgar F. (Ted) Codd (1923–2003) published a fundamental paper concerning a new model of organizing data in databases: *A Relational Model of Data for Large Shared Data Banks* first appeared in 1969 as an IBM Research Report with the title *Derivability, Redundancy, and Consistency of Relations Stored in Large Data Banks*. Codd was working at IBM but, at the beginning, the company did not realize the power and impact of the new approach. The model attracted theoretical studies for its elegance and formal crispness but for about ten years only experimental implementations (e.g. System R) were realized. Ted Codd received the Turing Award in 1981 "for his fundamental and continuing contributions to the theory and practice of database management systems, especially relational databases".

Fig. 5.2 Edgar F. (Ted) Codd

inally formulated in the unpublished paper *A Type-Theoretical Alternative to CUCH, ISWIM, OWHY*.

Finally, the two PhD theses by Jean-Marie Cadiou (*Recursive Definition of Partial Functions*) and Jean Vuillemin (*Proof Techniques for Recursive Programs*), based on the work of Dana Scott (but also on the work of John McCarthy), were submitted in Stanford in 1972 under the supervision of Zohar Manna. The theses concerned the fixpoint approach to the semantics of recursive programs, in which the two computation rules *call by name* and *call by value* are analyzed with respect to their power to capture the fix point of a recursive program, establishing a connection between the operational and the denotational semantics of programs.

From March 20 to 22, 1972, an important symposium on *Complexity of Computer Computations* was organized at the IBM Thomas J. Watson Research Center in Yorktown. Computational complexity was at that time one of the hot subjects in the theory of computing and six months before, at the Courant Institute of Mathematical Sciences in New York, another symposium had been held on this topic: the *Courant Computer Science Symposium 7: Computational Complexity* organized by Randall Rustin. The Courant symposium had been devoted to various subjects among which a major role was played by machine-independent axiomatic computational complexity. To this subject were addressed the talks of Patrick Fischer, Manuel Blum and John Gill, Robert Constable, Nancy Lynch, Martin Davis, and Paul Young. An extensive bibliography on computational complexity, consisting of almost 2000 items, had been provided by M. I. Irland and Patrick Fischer. The IBM symposium had an entirely different focus: analysis of algorithms and complexity of problems. At the symposium were present most of the scientists who had contributed to the development of efficient algorithms in various domains:

Michael Fischer, Robert Floyd, John Hopcroft, Richard Karp, Mike Paterson, Michael Rabin, Volker Strassen, Robert Tarjan. The relevance of that symposium is put in evidence in the introduction of the proceedings where the editors, Raymond Miller and Jim Thatcher, say: "The Symposium dealt with complexity studies closely related to how computations are actually performed on computers. Although this area of study has not yet found an appropriate or generally accepted name (*sic!*), the area is recognizable by the significant commonality in problems, approaches, and motivations. The area can be described and delineated by examples such as the following. (1) Determining lower bounds on the number of operations or steps required for computational solutions of specific problems such as matrix and polynomial calculations, sorting and other combinatorial problems, iterative computations, solving equations, and computer resource allocation. (2) Developing improved algorithms for the solution of such problems which provide good upper bounds on the number of required operations, along with experimental and theoretical evidence concerning the efficiency and numerical accuracy of those algorithms. (3) Studying the effects on the efficiency of computation brought about by variations in sequencing and the introduction of parallelism".

The proceedings of the symposium include what can be considered, probably, the most important paper that appeared in that year and one of the most important papers in the entire history of computer science: *Reducibility Among Combinatorial Problems*, by Dick Karp, actually the only paper in the symposium that was not devoted to new results in algorithmics but rather to computational complexity. This landmark paper, that by now counts more than 9,600 citations, has provided the building blocks that still today are at the base of complexity analysis of combinatorial problems. First of all the class **P**, containing the decision problems that can be solved in polynomial time by means of a (deterministic) Turing machine, is identified with the class of computationally tractable problems "a reasonable working hypothesis, championed originally by Jack Edmonds in connection with problems in graph theory and integer programming, and by now widely accepted". Then the class **NP** is defined as the class of problems that can be solved in polynomial time by means of a nondeterministic Turing machine and it is observed that many important problems for which we do not have polynomial-time algorithms indeed belong to the class **NP**.

In order to analyze the complexity of these problems the notion of *reducibility* between decision problems (later called *Karp reducibility*) is introduced: a fundamental tool that maps one problem into another by providing an algorithm which transforms any positive (negative) instance of the first problem into a positive (negative) instance of the other problem. If problem A can be transformed into problem B in polynomial time then problem B is at least as hard as problem A; moreover, if one might solve problem B in polynomial time then also problem A could be solved in polynomial time, by transforming, in polynomial time, any of its instances into an equivalent instance of B (which can be done due to the Karp reducibility) and then solving such instance in polynomial time.

Finally, the notion of **NP**-completeness is introduced. A problem A is defined by Karp as **NP**-complete if it belongs to **NP** and if the satisfiability problem (that is the problem to decide whether a given propositional formula is satisfiable) can be reduced to A[9].

Karp's work had been "stimulated" (as he says) by a fundamental result that appeared the year before. At the *Symposium on Theory of Computing* (STOC) 1971 Steve Cook had presented the paper *The Complexity of Theorem-Proving Procedures* (again a milestone in the history of computer science and, more generally, of mathematics, by now cited in more than 6,400 papers) in which he had shown that (in his words): "any recognition problem solved by a polynomial time bounded nondeterministic Turing machine can be 'reduced' to the problem of determining whether a given propositional formula is a tautology. Here 'reduced' means, roughly speaking, that the first problem can be solved deterministically in polynomial time provided an oracle is available for solving the second"[10]. Intuitively speaking, the result implied that the language consisting of propositional formulae that are tautologies, i.e. formulae that are satisfied by any assignment of truth values to their variables, has a very rich combinatorial structure, a very strong expressive power that makes the recognition of such languages harder than any problem that belongs to the class **NP**.

Let's see how Karp, in a later paper[11], explains how he was influenced by Cook's work: "It was not by accident that I was struck by the significance of Cook's Theorem. My work in combinatorial optimization had made me familiar with the traveling salesman problem, the maximum clique problem and other difficult combinatorial problems. Jack Edmonds had opened my eyes to the possibility that some of these problems might be intractable. I had read a paper by George Dantzig showing that several well-known problems could be represented as integer programming problems, and suggesting that integer programming might be a universal combinatorial optimization problem. My reading in recursion theory had made me aware of reducibilities as a tool for classifying computational problems".

Although the basic ingredients were already implicit in Cook's work the step forward realized by Dick Karp in the 1972 paper had an immense im-

---

[9] Note that in the first papers on the subject the term **NP**-complete is often replaced by the term **P**-complete. Today the two terms denote clearly different concepts: we define **NP**-complete as a decision problem that belongs to **NP** and has the property that all problems in **NP** can be reduced to it by means of polynomial time transformations (a definition that is equivalent to the one used by Karp) while we define **P**-complete a decision problem that belongs to **P** and has the property that all problems in **P** can be reduced to it by means of transformations using a logarithmic amount of space.

[10] We may see an oracle for a set S as a black box able to answer in constant time queries such as *Does x belong to S?*

[11] *The Mysteries of Algorithms* in *People and Ideas in Theoretical Computer Science*, C.S. Calude, Ed., Springer, 1998.

pact. By applying his definition of reducibility (somewhat stronger than the definition used by Cook), for the first time Karp was able to analyze and classify the complexity of 25 major combinatorial decision problems, showing that all of them were **NP**-complete, i.e. they had the same complexity as the satisfiability problem (modulo a polynomial-time transformation), the beginning of a long path that has led to the identification of several thousand **NP**-complete problems.

Again the words of Karp[12] explain the relevance of his contribution (reflecting a mild criticism of the preceding foundational research in computational complexity): "The early work on **NP**-completeness had the great advantage of putting computational complexity theory in touch with the real world by propagating to workers in many fields the fundamental idea that computational problems interesting to them may be intractable, and that the question of their intractability can be linked to central questions in complexity theory". The relevance of the paper (for which the author acknowledges the contributions of Gene Lawler and Bob Tarjan) and of its subsequent journal version[13] lies not only in the fact that the complexity of several important problems was characterized by showing that they are (as the satisfiability problem) the hardest problems belonging to the class **NP** but is also related to the fact that, in this very first paper on the subject, Karp also provides other characterizations of the class **NP** that can be expressed more naturally than resorting to an exotic computation model such as Turing machines.

In particular, he shows that the problems in **NP** are those problems for which to find a solution may be (as far as we know) exponentially 'hard' but to verify a solution is 'easy', i.e. doable in polynomial time. Note that since the classes **P** and **NP** have been introduced the question whether indeed problems with this characteristic exist (or, in other words, whether there exist any problems that belong to **NP** but not to **P**) is one of the major open problems in computer science and mathematics. In the year 2000 the Clay Mathematical Institute included the problem '**P** versus **NP**' as one of the seven Millennium Problems for whose solution a one million dollars prize has been offered.

Observe also that, by the polynomial-time reducibility relation holding between problems in **NP**, finding a polynomial algorithm for any **NP**-complete problem would imply that this would hold for all problems in the class **NP**: this would make a tremendous impact on most fields of science and engineering, since most optimization problems related for example to analysis of data and to planning would turn out to be solvable in polynomial time.

---

[12] Ibidem.

[13] The paper was published in journal version in the journal *Networks* three years later with the title *On the Computational Complexity of Combinatorial Problems*. It is interesting to note that Karp changed the list of **NP**-complete problems at the end of the paper. "Most of the problems in the original list I published in 1972 are superseded here by more specialized problems which have since been discovered to be **NP**-complete." So for example 0-1 integer programming is eliminated and the planar degree constrained 3-coloring is included.

Fig. 5.3 Larry Stockmeyer

Another major paper that is a pillar of computational complexity and that again appeared in this fantastic year, 1972, is *The Equivalence Problem for Regular Expressions with Squaring Requires Exponential Space* by Albert Meyer and his student Larry Stockmeyer (Fig. 5.3), presented at the *Symposium on Switching and Automata Theory* (SWAT). In this paper Meyer and Stockmeyer introduced the polynomial-time hierarchy[14], a fundamental step that completed the construction initiated by Steve Cook. The polynomial-time hierarchy is an infinite hierarchy of families of languages (in which **P** is at level 0 and **NP** is at level 1) that allow us to characterize the complexity of all problems that can be described by a sequence of alternating quantifiers followed by a polynomial-time decidable predicate. For example, it is easy to see that the set of propositional formulae of length k such that there exists a shorter equivalent formula belongs to the second level of the hierarchy; in fact the set can be expressed as follows: given a formula φ and an integer k there exists a formula φ' such that the length of φ' is smaller than k and for all truth assignments φ is true iff φ' is true.

But Cook, Karp, Meyer were not the only researchers who were studying the characterization of the complexity of combinatorial problems in those years. Meanwhile, in a much different world, Leonid Levin (Fig. 5.4) was thinking about the same issues, but not getting nearly the same publicity. In the Soviet Union, at that time, many researchers were considering questions related to the **P** versus **NP** question. In particular there was the notion of the class of problems that could only be solved by *perebor*, the Russian name for algorithms that were essentially based on exhaustive search (that is, by essentially considering all possible solutions to the problem and checking

---

[14] The polynomial-time hierarchy was later systematically analyzed by Larry Stockmeyer who illustrated all its characteristics and properties in another important paper (that now has 1330 citations): *The Polynomial Time Hierarchy*, published in Theoretical Computer Science in 1976. Larry died prematurely in 2004, when he was only 56.

Fig. 5.4 Leonid Levin

whether they are indeed solutions)[15]. In April 1972, in Novosibirsk, Leonid Levin presented his work (published in 1973 with the title *Universal Sequential Search Problems*) in which, independently from Cook, he had essentially come to the same result: he had discovered that the search version of the satisfiability problem[16] was 'universal' in the sense that all **NP** search problems could be reduced to it.

For me 1972 started with the participation (for the first time as lecturer) in a school in Denmark. The school was organized by Brian H. Mayoh (Fig. 5.5), a British professor who had been hired by Aarhus University in 1965, just after he had finished his PhD in logic from the University of Illinois at Urbana-Champaign. He was associated with the Institute of Mathematics, and appointed as one of the main players in building up activities in com-

Fig. 5.5 Brian Mayoh

---

[15] David S. Johnson, *A Brief History of* **NP**-*Completeness*, 1954–2012, *Documenta Mathematica* - Extra Volume ISMP (2012)

[16] The search version of the satisfiability problem consists in the following: "given a propositional formula, find a satisfying truth assignment", while in the corresponding decision problem we ask: "given a formula does there exist a satisfying truth assignment?".

puter science, supplementing existing activities in mathematics and statistics at the institute. Brian contributed a lot to make Aarhus the capital of theoretical computer science in Denmark, a role that Aarhus has maintained until now. The school had a strange title: *Open House in Unusual Automata Theory*. Brian's intention in organizing the school was his vision that Automata Theory and Formal Languages had much more to offer besides the classical applications at the time in, for example, the syntax of programming languages and compiler construction. Indeed at the school various new topics in the field of formal languages and automata theory were addressed: stochastic automata (François Bancilhon, Azaria Paz), cellular automata (Gabor Herman, Paul Vitanyi), Lindenmayer's developmental systems (also known as *L*-systems: Gabor Herman, Arto Salomaa, Paul Vitanyi and Aristid Lindenmayer himself). Finally, I lectured about various approaches to computational complexity (subrecursive hierarchies, complexity of Turing machine computations and abstract computational complexity).

Taking part in that school, interacting with outstanding researchers such as Arto Salomaa, Azaria Paz, and Paul Vitanyi, opening my eyes to new research domains, was a wonderful experience. *L*-systems in particular, to which various lectures at the school were devoted, sparked my interest. Developmental systems had been introduced by Lindenmayer in 1968 (in the paper *Mathematical Models for Cellular Interaction in Development* published in the *Journal of Theoretical Biology*) in order to model morphogenetic processes in growing organisms. In a sense, an *L*-system could be seen as an array of automata where a single automaton could duplicate itself or die on the basis of messages coming from neighbor automata. The new computational mechanism that remarkably contained in itself an intrinsic notion of parallelism and that allowed us to characterize classes of languages different from classical Chomsky languages appeared to me very fascinating (especially in the animated version provided by CELIA, a software system realized at SUNY in Buffalo by Gabor Herman and Adrian Walker). For the first time I learned how developmental systems could solve the Dutch national flag problem (that consists in sorting items of three different colors in the correct order), how they could simulate a worm that reproduces head and tail after being cut, and how they could help to understand other developmental phenomena of biological nature such as the growth of leaves in a plant according to a Fibonacci-like scheme.

Among the other participants John Myhill lectured about intuitionistic set theory, Bernd Reusch about linear automata, Jan van Leeuwen (still a graduate student at that time) about generalizations of context-free languages. The school also gave me the chance to open a window on the Scandinavian world. The clean, productive and well organized atmosphere of the Department of Computer Science on Ny Munkegade stimulated at the same time concentration and socialization. This warm feeling started from the very first day of my arrival, when a beer in a pub with two young researchers, Mogens

Nielsen and Erik Meinecke Schmidt, signed the kick off of a long friendship, and persisted on all occasions when I had chance to return to Aarhus.

Back from Aarhus two meetings were waiting for me. On January 24 the promoters of GRIT met at Milan Polytechnic. The meeting had mostly the aim to do a survey of the researchers involved in theoretical informatics and of the research subjects that were addressed in the various Italian sites, both in view of the creation of the Italian group and of the meeting to be held soon in Brussels. The result of the census was that about 60 researchers were involved altogether. Then it was proposed that the promoters would again meet on February 2 in Pisa with the aim to take final decisions regarding the creation of GRIT.

Then on January 27 and 28, 1972 the meeting in Brussels came. Let's see how Maurice Nivat himself reported about the meeting. "C'est en Janvier 1972 à Bruxelles, dans les bâtiments de la Commission des Communautés Européennes, et à l'instigation de celle ci que se sont réunis un certain nombre d'informaticiens, connus pour la nature assez théorique de leurs travaux. M. Caracciolo présidait cette réunion. Un texte élaboré par les membres de l'Université de Paris VII présents à cette réunion fut présenté à cette occasion pour tenter de définir ce que l'on peut appeler l'Informatique Théorique, ce chapitre de l'Informatique qui consiste à utiliser tous les outils mathématiques et logiques pour préciser et étudier la notion de calcul. Les principaux sous-chapitres, mais cette liste n'est pas limitative au contraire, en seraient: (i) la théorie des algorithmes et de leur complexité, (ii) la théorie des automates et des langages formels, (iii) la théorie de la programmation (sémantique formelle des langages de programmation). En même temps fut proposé la création d'une Association Européenne qui permettrait de favoriser le développement de cette discipline"[17].

Researchers from six countries plus several officers of the European Commission took part in the meeting. France was represented by Maurice Nivat, Louis Nolin, and the linguist Maurice Gross; Germany by Hans Langmaack and Karl Heinz Böhling, The Netherlands by Leo Verbeek and Jaco de Bakker, the UK by Mike Paterson, Belgium by Michael Sintzoff (Fig. 5.6), Italy by Corrado Böhm, Ugo Montanari and Giorgio Ausiello. The meeting was chaired by Alfonso Caracciolo. The agenda of the meeting included a survey of the activity of the Group PREST, the presentation by Louis Nolin of the document containing the definition of the field of 'informatique théorique' and an overview of the importance of the field, an examination of the situation in the different countries, and finally the presentation, discussion and possibly approval of the proposal prepared by Maurice Nivat on the cooperation among universities. This was indeed the hottest issue. At first the idea was to follow the EMBO example and to ask for a substantial funding of European institutions.

---

[17] This report is contained in a history of the life of EATCS during its first 25 years written by Ute and Wilfried Brauer on the occasion of the EATCS Silver Jubilee, that appeared in Issue 62 of the Bulletin of EATCS (June 1997).

Fig. 5.6 Michel Sintzoff

At the meeting a list of the costs required to run the association and to perform the expected activities in the first year was formulated: 50,000 USD for visits and exchanges of professors and researchers, 30,000 USD for one symposium, for six meetings and for starting a journal (an idea that was already present in the mind of Maurice Nivat but that became a reality only three years later). So 80,000 USD total. The requested funding would increase to about 150,000 – 200,000 USD in the subsequent years. The view of other participants was instead in favor of a lighter form of cooperation. A passionate discussion followed and, although the precise form that the initiative would take was still unclear, in any case the decision to start some kind of European cooperation in theoretical computer science was taken. Michael Sintzoff was asked to act as president in the preparation of all documents that were needed to formalize the creation of the association (scientific presentation in line with Nivat's document on 'Informatique théorique' and two supplementary documents presented by Böhm and de Bakker, statute, proposed budget, etc.).

Michael Sintzoff wholeheartedly started the coordination work asking us for comments and refinements to the original documents. A series of meetings (on March 10 and on March 29 in conjunction with a meeting of representatives of universities) were held in Brussels, still under the sponsorship of the 'Group de formation en Informatique' chaired by Desfosses. Finally the European Commission gave its *placet*: go ahead!

After not so long it became clear that the EMBO experience could not be followed. This is also reflected in the changes in the title of the initiative and in the articles of the proposed statute. At the beginning a document dated February 7th has the title *Institut Européen d'Informatique Théorique* and considers that the institute should consist of 'associated institutes' ("universités, instituts, laboratoires ayant des activités de recherche et d'enseignement en informatique théorique") and of simple 'members' ("personnes physiques s'occupant d'informatique théorique ou de disciplines qui s'y rattachent"). Subsequently the proposal had been transformed into *As-*

sociation Européenne d'Informatique Théorique – AEIT (in English *European Association for Theoretical Computer Science* – EATCS), and while still open to the participation of institutional members it had been conceived essentially as an association of 'individual members'. The aim of the AEIT was to promote the "development of research and education in theoretical computer science in Europe". This aim would be pursued by means of exchanges of professors, researchers and students, organization of meetings and conferences, dissemination of research results, organization and coordination of joint research projects, and organization of research activities jointly with persons and institutions not belonging to the AEIT.

Various provisional drafts of the statutes of the association prepared by EEC legal experts were circulated. On June 3rd Michael Sintzoff met with de Bakker, Nivat and Paterson and they prepared the final document for the creation of EATCS (see Appendix D). The document would carry the names of the founders: Giorgio Ausiello (Italy), Jaco de Bakker (The Netherlands), Maurice Nivat (France), Mike Paterson (UK), Manfred Paul (Germany, rather, to be more precise, the Federal Republic of Germany), Michel Sintzoff (Belgium), and Leo Verbeek (The Netherlands). In the same meeting it was decided to nominate Leo Verbeek as president, Manfred Paul and Mike Paterson as vice-presidents, Maurice Nivat as secretary and Michael Sintzoff as treasurer. In order to be recognized at the European level the association had to be created according to Belgian laws (this is why the original statute is in French) and would have its legal address in Belgium (indeed this was decided to be in the house of the pro tempore treasurer Michel Sintzoff: Avenue du Château 7, Rixensart). On June 24 the document was submitted to the Belgian authorities and on September 4, 1972 the King of the Belgians, Baudouin, accorded the status of legal person to EATCS. Finally, on October 2 a letter from the Belgian Ministry of Justice was sent to Michael Sintzoff with a copy of the 'arrêté royal du 4 septembre 1972 qui accorde la personnalité civile à l'Association Européenne d'Informatique Théorique et qui approuve les statuts de cette association internationale' and with the information that 'L'arrêté et les statuts précités ont été publiés par mention au Moniteur belge du 26 septembre 1972.' EATCS was officially born.

Of course this was just the first step. The real developments had still to be invented and in order to give the association the role that we had sought the first step would be to enlarge the membership and to solicit the involvement of the most prominent European theoretical computer scientists. On September 6 Maurice Nivat wrote a letter to the founders of EATCS[18] calling for the first meeting of the council and the first general assembly of the association, to be held in Warwick in Spring 1973 and suggesting the text of a letter to be sent to a large group of European colleagues to ask them to join the newborn association. After resuming the history of the foundation of the association and illustrating the composition of the council the letter said: "Il

---

[18] Actually the letter carried the signature of J.-F. Perrot due to the fact that Maurice Nivat was 'empeché'.

importe donc beaucoup, cher collègue, que vous acceptiez de grossir les rangs des premiers membres de cette association et que, à l'egal de ses fondateurs, vous contribuiez non seulement à la faire vivre mais à en définir la composition, les buts, les moyens d'action. Le bureau provisoire attend de vous, au delà d'une simple adhésion, votre concours actif." The letter contained a list of colleagues that, in Maurice's view, should join the association so that it could play an effective role: Böhm, Caracciolo, Jacopini, Grasselli, Montanari (for Italy[19]), Schnorr, Brauer, Hotz, Schönage, Langmaak, Eickel, Boehling (for West Germany), Schützenberger, Perrot, Berstel, Kahn, Nolin, Gross (for France), Engeler and Strassen (for Switzerland), Scott, Park, Cooper, Burstall, Milner, Hoare, Strachey, Landin (for the UK), Dijkstra, Rozenberg, De Bruijn (for The Netherlands), Rabin, Shamir, Paz, Cohen, Manna, Ginzburg (for Israel, whose adhesion to the association had been considered essential since the beginning) and, furthermore, Mayoh (Denmark), Salomaa (Finland), Aanderaa (Norway), Bekic (Austria).

The invitation letter (see Appendix E) was sent in the subsequent months and most of the colleagues invited indeed joined the association. Of course also various US names had been circulated (although not mentioned in Maurice's letter) and were invited to join, one above all, Ronald Book, one of the most prominent researchers in formal languages. Finally it was decided to invite, in a second phase, colleagues from Eastern Europe. Everything was ready for the first general assembly.

But let us go back to January 1972. After the meeting in Brussels, on the way back to Rome I made a stop in Paris. At that moment Paris was indeed the capital of European 'informatique théorique' and our French colleagues were actively promoting initiatives of various kind also thanks to the organizational support of IRIA. In particular Maurice Nivat and Jean-François Perrot (Fig. 5.7) had started a regular seminar under the title: 'Seminaires IRIA - Théorie des algorithmes, des langages et de la programmation'.

Fig. 5.7 Jean-François Perrot

[19] In my reply I suggested to Maurice to include in his invitation list three more Italian colleagues: Aldo De Luca, Giuseppe Trautteur and Gianni Degli Antoni.

After the 1968 turmoil the University of Paris had been divided into several smaller universities. The organizers belonged to the University of Paris VII *Denis Diderot*, but the lectures were taking place in the historical building of La Sorbonne, probably because the Jussieu buildings were still under construction. Some of the talks, collected in a booklet printed by IRIA, give an idea of the content of the seminar. Giorgio Germano, from Naples, was invited to present his joint work with Andrea Maggiolo on a new notion of computability based on Markov algorithms; Klaus Indermark presented a paper in which he showed that push down transducers belong to the (word function) Grzegorczyk class $\mathcal{E}_1$; Joost Engelfriet, from Twente, spoke about the relationship between program schemes (a syntactic abstraction of the notion of program[20]) and formal languages, a subject particularly popular in that period and rather close to the approach to semantics of programs proposed by Nivat; finally Paul de Roever, from Amsterdam, presented the work he had done under the supervision of Jaco de Bakker on computation rules in recursive program schemes.

For me it was undoubtedly a great honor to be invited to give a talk in the seminar. Jean-François Perrot introduced me to the audience at La Sorbonne. It was my first talk in French! I gave a survey of various approaches to computational complexity, trying to show the relationship among various subrecursive hierarchies (especially on the basis of the latest papers by Dennis Tsichritzis and by Gabor Herman on this subject), and the connections among subrecursive hierarchies, Turing machine complexity and abstract computational complexity. At the end two cheerful young smiling guys came to talk to me; one had brown curly hair, the other was blond with a Flemish look: they were Philippe Flajolet and Jean-Marc Steyaert, both at that time PhD students at University Denis Diderot under the supervision of Maurice Nivat; they were among the first French computer scientists working in computational complexity. They briefly illustrated their work to me and we discussed their possible visit in Rome. Again a deep, longlasting friendship started at that moment.

The foundation of EATCS immediately stimulated the interest of the researchers in all European countries and still today (more than 40 years later) it can be considered a big success for having fostered the spirit of a large European scientific community (although it has to be noted that soon the association started to involve also a large number of American and Asian scientists). In 1972, together with other initiatives that were taken in various countries in that period, the foundation of EATCS contributed to give an identity to the researchers working in theoretical computer science and to turn on the light on the mathematical foundations inside a field, informatics, that was dominated by the continuously evolving technological drives. The reason for the success was also due to the fact that in that period no

---

[20] The concept of program scheme (in the monadic case) and its relationship with finite state automata had been originally studied by Ianov in 1960 in his paper *On the Logical Schemes of Algorithms*.

similar institution had yet been created at national level in Europe. Only in the United States two organizations existed and, in a sense, were a reference model for European projects: SIGACT (the *ACM Special Interest Group on Automata and Computability Theory*[21]), founded in 1968, and TCMFC (the *IEEE Technical Committee on Mathematical Foundations of Computing*), active since 1960 when the first *Symposium on Switching Circuit Theory and Logical Design* was held[22]. The main difference was that such organizations could rely on support from their mother societies while the European association would start its activity without any financial basis.

In Italy, in particular, the foundation of EATCS acted as a catalyst to push further the activity of GRIT. Two meetings were held in Pisa on February 2 (when the group was formally approved) and in Rome on February 14. Corrado Böhm was nominated coordinator of GRIT and Ugo Montanari secretary. It was also decided to organize a conference devoted to theoretical informatics in Pisa in 1973 and to print a regular bulletin for information dissemination on scientific events, visits of foreign scientists, publications, etc. At the same time another more ambitious project was initiated by the same group of researchers, the creation of a much larger group including all Italian researchers in informatics. This group, to be called GRI, was aimed on one side at providing an alternative to Ruberti's project to gather Italian computer scientists under the umbrella of system theory and, on the other side, at involving computer engineers who, at that time, were organized in a joint group with electrical engineers. Like similar organizations, GRI would be structured with special interest subgroups. In this context GRIT (the group of theoretical computer scientists) would be simply a subgroup of GRI. In a sense theoreticians were now leading a big effort to assemble all Italian computer scientists, overcoming previous divisions originated by belonging to different local 'schools' or to different cultural cradles (mathematicians, physicists, electrical engineers).

On March 15, finally, a big meeting of all Italian computer scientists was held in Pisa. More than 130 researchers got together in the big conference room of IEI in Via Santa Maria. At the end of the meeting the statutes of the new group were defined and Bruno Fadini was elected president. At the end of a second meeting, held in Bologna on April 27, GRI could already count on 178 members belonging to universities and CNR research centers from all over Italy. It is remarkable that 64 out of 178 GRI members, that is more than one third, were in fact people working in theoretical informatics and interested to be included in the GRIT subgroup. The proposed subgroups, approved in a subsequent meeting in Bologna on July 10–11, were the following ones: theoretical informatics (coordinated by Giorgio Ausiello), theoretical founda-

---

[21] The name of SIGACT was later changed to *Special Interest Group on Algorithms and Computation Theory* for a better correspondence with the main focus of the research interests of the group members.

[22] As we already said, subsequently the name of the symposium became *Switching and Automata Theory* and finally *Foundations of Computer Science*.

tions of operating systems (Norma Lijtmaer), computer networks (Gesualdo Le Moli), information storage, management and search (Giampio Bracchi), computer assisted instruction (Alfio Andronico), pattern recognition (Stefano Levialdi), and mathematical programming (Francesco Maffioli).

Soon GRI's president established contacts with the Special Commission for Informatics of CNR (whose president was Antonio Borsellino, a physicist expert of quantum electrodynamics but also interested in cybernetics, and whose secretary was Arrigo Frisiani, professor of electronic computers, both from Genoa) and submitted a series of requests, in particular the request to define within CNR a unique entity able to provide the necessary funding for computer science research. Unfortunately such a big effort, although very important for helping the growth of the Italian computer science community and for establishing a leading role for theoreticians in this community, did not have any direct impact on the funding institutions. The Special Commission for Informatics proposed to CNR authorities the creation of a formal group for informatics and, consequently, the provision of specific funds for research activities in the field, but this attempt did not produce any result and we had to wait until 1979 to see the creation of CNR initiatives specifically devoted to the promotion and funding of informatics.

Besides France and Italy also in The Netherlands the promotion of activity in the theory of computing was lively. In June an *Advanced Course on Programming Languages and Data Structures* was organized by van Wijngaarden and was held at the University of Amsterdam. Beside van Wijngaarden himself the lecturers included two pioneers of computer engineering: Maurice Wilkes from Cambridge, famous for having built one of the first programmable computers, the EDSAC machine; Willem van der Poel from Delft, who contributed to the development of Algol 68 and of LISP for the early Dutch computers; and Dana Scott from Princeton, whose work on semantics of programs we have already mentioned. Also in Germany theoretical computer science was gaining momentum. Inside the German society for informatics (Gesellschaft für Informatik), founded in 1969, the special group on Theory of Automata and Formal Languages (*Automatentheorie und Formale Sprachen*) had been recently created.

For another reason 1972 is a landmark year in the history of theoretical computer science: in that year two important series of European conferences started, ICALP and MFCS, and several other events (beside those we have already mentioned), marked the interest in theoretical computer science that was increasing around the world. From the 3rd to the 7th of July, the *Colloque sur la Théorie des Automates, des Langages et de la Programmation* was held in Rocquencourt, near Paris, in the premises of IRIA (see Appendix F). The conference was organized under the sponsorship of ACM-SIGACT and although the newly born EATCS was not formally involved in this first edition, the conference would become (with the name *International Colloquium on Automata, Languages and Programming*) the permanent annual conference of the European association[23]. For the first time in Europe a conference was devoted to all aspects of theoretical computer science and could compete with the recently born *Symposium on Theory of Computing* (STOC). The Program Committee consisted of outstanding European and US researchers, most of whom were present at the conference: Corrado Böhm, Samuel Eilenberg, Pat Fischer, Seymour Ginzburg, Gunther Hotz, Michael Rabin, Arto Salomaa, Adriaan van Wijngaarden. It was chaired by the father of the French school of theoretical computer science: Marcel-Paul Schützenberger.

The conference was a big success. The program was excellent and the sessions corresponded to the subdomains defined in Nivat's document on 'informatique théorique': Théorie des automates, Théorie des langages, Complexité des algorithmes, Théorie de la programmation. Beside a few senior scientists such as Ron Book, Joost Engelfriet, Michael Harrison, Zohar Manna, Louis Nolin, Eli Shamir, a great number of young researchers who would later make the history of theoretical computer science presented their papers: Alberto Bertoni, Jaco de Bakker and William Paul de Roever, Robert Cori, Stefano Crespi Reghizzi, Philippe Flajolet and Jean-Marc Steyaert, Burkhard Monien, Jean Vuillemin, and many others. Out of the 45 papers selected (at the conference the papers could be submitted in German, English and French, and actually two of the papers were in German and 14 in French) the largest part (23) were devoted to automata and formal languages, while 12 papers were devoted to theory of programming and 11 to computational complexity. Taking into account that many papers in the field of complexity were indeed referring to formal languages, it is clear that syntactic aspects regarding languages and automata were still dominating the scene of theoretical computer science (see Appendix G).

Semantics was in any case gaining attention. An important paper concerning the *Fixpoint Approach to the Theory of Computation* (essentially based on Vuillemin's PhD dissertation) was presented by Manna and Vuillemin, while Maurice Nivat proposed his syntactic approach to semantics in the paper *Langages algébriques sur un magma libre et sémantique des schémas de programme*.

Although the call for papers identified four major domains: theory of automata and formal languages, theory of algorithms and complexity, formalization of the notion of program and machine, and semantics of programming languages, only some of the topics were addressed. Very few contributions were devoted to novel approaches to the notion of program and machine. No contribution was devoted to the presentation of new algorithmic results; the spectrum of addressed topics would deeply change in subsequent years[24].

---

[23] Since the second edition the sponsorship of the EATCS has been acknowledged.

[24] Algorithms are nowadays by far the most popular field in a typical ICALP program. In the program of ICALP 2016, for example, 89 papers belonged to Track A (Algorithms, Complexity and Games), 34 to Track B (Logic, Semantics, Automata and Theory of Programming) and 23 to Track C (Foundations of Networked Computation: Models, Algorithms and Information Management).

Fig. 5.8 Mogens Nielsen

During the conference Mogens Nielsen[25] (Fig. 5.8) approached me and, on behalf of Brian Mayoh, invited me to join the crew in Aarhus for one semester. The invitation was very appealing but again I declined. A few days later the birth of my second child, Irene, was expected.

Just one month later, the Polish computer scientists organized another conference destined to become the second major European event devoted to theory of computing: the *Symposium on Mathematical Foundations of Computer Science* (MFCS). The symposium had been preceded by another important event in Eastern European countries: the *International Symposium on Theoretical Programming* that had been organized in Novosibirsk, known at that time as the Soviet science park, by Andrey Ershov[26] and Valery Aleksandrovic Nepomniaschy. The symposium was probably the first occasion for a big gathering of Eastern and Western computer scientists in the Soviet Union and was a great success. The proceedings (published by Springer two years later) show the participation of several European and US scientists (Hoare, Paterson, Milner, Engeler among others) together with several important figures of the Russian computer science scene (Barzdin, Trakhtenbrot[27], Kotov, etc.).

The first edition of MFCS, which actually consisted of both a symposium and a summer school[28], was organized by the Computation Centre of the

---

[25] Mogens Nielsen (born in 1949) obtained his Master under the supervision of Brian Mayoh and his PhD under the supervision of Arto Salomaa. Mogens has been one of the leaders of European research in program semantics. From 1993 to 1997 he was co-Director of the Danish Basic Research Institute for Computer Science (BRICS) and from 2002 to 2006 he was President of EATCS.

[26] Andrey Ershov (1931–1988) was one of the pioneers of Soviet computer science in the field of programming language research.

[27] Boris Trakhtenbrot (1921–2016) was one of the greatest Soviet mathematicians and logicians. He was among the first scientists who in the 1960s addressed important issues in the theory of computing and who proved relevant results in automata theory, formal languages and computational complexity, sometimes anticipating the work done by Western theoreticians. In 1980 he moved to Israel where he became professor at Tel Aviv University.

[28] Marginal but cute: the registration fee for the first MFCS was 20 USD!

Polish Academy of Sciences and by the Institute of Computing Machines
of Warsaw University, and was held in Jablonna, 15 km from Warsaw[29] in
August 1972. The chairman of the first MFCS was Andrzej Blikle and the Organizing Committee included the *Gotha* of Polish mathematical logic and theoretical computer science: Helena Rasiowa[30] (1917–1994), Zdzislaw Pawlak[31]
(1926–2006), Wladislaw Turski (1938–2013), Antoni Mazurkiewicz, Andrzej
Salwicki, etc. The call for papers (see Appendix H) had a curious definition
of the scope of the conference: "Because of a large scope of this subject it is
not possible to deal with all its branches during a one week meeting. For this
reason the organizers intend to limit this scope to problems that are developed in Poland, i.e. to semantics of programming languages and foundations
of computing systems. In particular the following topics will be dealt with:
proving properties of programs; correctness, equivalence, adequacy and related topics; program transformations; primitives of programming languages
and computing systems".

The Fall of 1972 was still rich in events. In particular at the end of November the first meeting on complexity theory was held in Oberwolfach, at the
Mathematisches Forschungsinstitut. The institute, founded in 1944 and run
since 1959 by the German Mathematical Society, had recently started to host
seminars also in theoretical areas of computer science, such as formal languages and automata theory. From November 26 to November 30, 1972 Claus
Schnorr, from Frankfurt, and Arnold Schönage, from Tübingen, organized
for the first time a *Symposium on Theory of Algorithms and Complexity*. A
large number of participants, mostly from Europe, were present. Among them
Bruno Buchberger, Philippe Flajolet, Jean-Marc Steyaert, Klaus Weihrauch
and László Kalmár[32]. Beside myself, the Italian participants were Giuseppe
Trautteur and Eliana Minicozzi.

Abstract computational complexity was still quite popular: Jiří Becvar,
Bob Daley, Peter van Emde Boas and I gave talks related to this area. But the
founding fathers of the field, Manuel Blum and Albert Meyer, had changed
their interest and it is worth mentioning their talks. Albert Meyer reported
on the work he had recently done with Mike Fischer and Larry Stockmeyer on
the inherent complexity of decidable problems in logics and automata theory
(such as for example the first order theory of a single monadic function or the

---

[29] Since the second edition that was held in Strbske Pleso (High Tatras, Czechoslovakia)
in 1973 it was decided that the conference would be organized alternatively, one year in
Poland and one year in Czechoslovakia.

[30] Helena Rasiowa was a great Polish mathematician. She was a student of Andrzej
Mostowski and made important contributions in logics, algebra and theoretical computer
science.

[31] Pawlak was the founder of the Polish school of artificial intelligence and one of the pioneers in computer engineering and computer science. He is famous for many contributions
to information retrieval and to artificial intelligence and in particular for having invented
the theory of rough sets.

[32] The logician László Kalmár (1905–1976) is considered the father of theoretical computer
science in Hungary and is known, in particular, for his work on elementary functions.

recognition of star-free regular expressions which define the empty language), work that is the basis of the famous paper that Michael Rabin presented at the IFIP World Congress in 1974 on *Theoretical Impediments to Artificial Intelligence*. Manuel Blum, instead, spoke about the work he was carrying out with his wife Lenore in Naples, where they had been invited by Trautteur and Minicozzi to spend their sabbatical: a recursion theoretic approach to inductive inferences, that is to the problem of identifying a program for a function f given a sequence of pairs $\langle x, f(x) \rangle$. Their work was inspired by the work of Ray Solomonoff (*A Formal Theory of Inductive Inference*, published in 1964) and of Marc Gold (*Language Identification in the Limit*, published in 1965), who showed how to build an inductive inference machine that can identify the class of primitive recursive functions (while no machine can identify all recursive functions). In the framework of inductive inference, by identifying a function f one means being able to derive, by means of the application of a suitable general algorithm to a collection of examples (argument value x, function value $f(x)$), an algorithm which is able, at least approximately, in some sense, to predict $f(z)$ for any new argument z. In this case, borrowing from Aristotelian terminology, we say that we induce a general rule (function f) from particular cases (the examples). Deriving general rules from a set of examples, in order to apply them for prediction, is a fundamental task in fields such as machine learning and artificial intelligence. At the foundational level, it is then important to be able to characterize functions on the basis of how "easy" it is to learn them (in the above terms).

Manuel and Lenore's main result consisted in showing how one can build a machine that identifies the class of 'compressed functions', i.e. all functions having optimal or near optimal programs. This work was an important step forward in the field and it was presented at SWAT in 1973. Subsequently it appeared in *Information and Control* in 1975 with the title *Toward a Mathematical Theory of Inductive Inference*.

Claus Schnorr, one of the organizers of the symposium, who had made previous contributions in abstract computational complexity, also gave a talk on a relatively new topic that was gaining attention from theoreticians: Kolmogorov complexity, that is the theory in which the complexity of a string (or of any other information structure) corresponds to the length of the shortest program (e.g., description of a Turing machine) that generates this string. The relation of this concept with inductive inference is clear and in fact the first contributions in this area are considered Solomonoff's papers: *A Preliminary Report on a General Theory of Inductive Inference*, from 1960, and *A Formal Theory of Inductive Inference* from 1964. Independently, in 1965, Andrey Kolmogorov (Fig. 5.9) introduced the notion of string complexity that takes his name, in the paper *Three Approaches to the Quantitative Definition of Information*, soon followed in 1966 by the related work of Gregory Chaitin *On the Simplicity and Speed of Programs for Computing Infinite Sets of Natural Numbers*. Important in this framework is the definition of the notion

Fig. 5.9 Andrey Kolmogorov

of a random sequence, and in various talks the connections between random sequences, Kolmogorov complexity and abstract complexity were addressed.

The symposium also included several talks devoted to algorithm design: Mike Paterson reported on on-line multiplication of integers and reals, Mike Fisher on work done in cooperation with Paterson on on-line pattern matching, Schönage on complexity of matrix transformations. Finally, various topics not precisely fitting with the title of the symposium were addressed. In particular Nivat and Müller presented papers dealing with the problem of translating recursive program schemes into iterative schemes. Remarkably, no **NP**-completeness results were yet presented at the symposium.

An interesting outline of the meeting in Oberwolfach was included in the third issue of the Bulletin[33] of GRIT, the Research Group on Theoretical Informatics, which in the meantime had become a subgroup affiliated to the Italian Research Group on Informatics, GRI. In a meeting held in Pisa on June 27, 1972 Montanari had resigned as GRIT secretary and I was asked to take his place and to take care of the editing of the Bulletin. Beside the report on the Oberwolfach Symposium the third issue, which appeared at the end of 1972, contained the minutes of the GRIT meeting held in Rome on November 17, an announcement of the second MFCS to be held in September 1973 in the High Tatras, a list of publications of the Turin group, and the results of the census taken in the previous February and regarding research activity in theoretical computer science throughout Italy.

According to the census there were 25 institutions where, at least partially, research in the theory of computing was carried out, among them 16 univer-

---

[33] The decision to print and distribute a Bulletin of GRIT had been taken in the GRIT meeting on February 14, 1972 and the second issue, printed in July, already contained relevant information for Italian theoreticians including the announcement of the *First Congress on Theoretical Informatics* to be held in Pisa in 1973 and the presentation of a computer bibliography on context free languages that had been realized in Rome.

sity departments (Physics, Mathematics, Information Science, Electronics, Automation), 6 CNR institutes and a few computing centers. Altogether the activities involved about 70 researchers. The survey also included a list of topics of interest that had been identified by the researchers. This survey provides an interesting overview of the focus of Italian theoretical computer science research in the early 1970s. The field that appeared to attract the most attention from researchers was the theory of programs. Here the addressed topics were

1. various approaches to semantics: operational (following the direction of the Vienna Method), mathematical (following the direction of Scott and Strachey), fix-point approaches (à la Vuillemin), other formal methods based on lambda-calculus and combinatory logic, axiomatic approaches based on higher order logics (following Park, Milner, de Bakker);
2. analysis of program schemes: syntactic approaches (à la Nivat), computational power of program schemes and translation from recursive to iterative program schemes, parallel program schemes (in the direction of the 1969 Karp and Miller paper, *Parallel Program Schemata*);
3. methods for proving program properties: termination, correctness, equivalence, properties of structured programs (according to Dijkstra and Wirth), etc.

In second place, as regards the number of researchers involved, came artificial intelligence and various topics in non-numeric computing: automatic theorem proving, problem solving (in the direction of Hewitt's PLANNER), heuristics (à la Nilsson), program synthesis, algebraic manipulation systems, natural language understanding, and question answering systems.

In the field of automata and formal languages the attention was concentrated on space and time bounded automata, probabilistic automata, fuzzy languages, cellular automata, algebraic characterization of abstract families of languages (à la Greibach), syntax analysis of context free languages, and grammatical inference.

Finally, a few areas were identified as being at the interfaces between theoretical computer science and other areas of computer science. At the interface with discrete mathematics, recursiveness and information theory, the chosen topics were analysis of algorithms (à la Knuth), identification of complexity lower bounds, reducibility between problems (à la Karp), abstract complexity (à la Blum), and Kolmogorov complexity. At the interface with operating systems was the whole area regarding properties of concurrent programs (deadlock prevention, liveness, determinacy, etc.). At the interface with information retrieval was the theory of information structures (data encodings, data structures, Codd's relational approach).

In Rome, our group at IAC, despite uneasy relations with the director, continued its activity. Marisa Venturini Zilli's research work concentrated on uniformly reflexive structures, an abstract, axiomatic approach to computability that had been introduced in 1969 by Eric Wagner and that al-

lowed us to extend properties of recursive functions (e.g., Kleene's recursion
theorem) to a large class of computation models. She was also engaged in
automatic theorem proving with Giuseppe Longo and in particular in the
quest for an efficient algorithm for unification of terms, an important and
intriguing problem that found a final solution in 1976 with the results of
Martelli and Montanari, and of Paterson and Wegman who independently
discovered a linear time algorithm. Puccio Jacopini continued his work on
models of λ-calculus while I was writing my lecture notes on formal languages
and computability, a text book for my students at the *School on Automatic
Control and Computing Systems*. We also had various visitors. After Andrzej
Grzegorczyck who spent three months with us lecturing on model theory, we
had the visit of Maurice Nivat and Louis Nolin and several other colleagues,
including Manuel and Lenore Blum who came to Rome during their stay in
Naples.

In that period theoretical informatics was gaining some attention also by
mathematicians. Encouraged by the contacts that some of them had with
French colleagues a meeting devoted to this discipline was organized by the
Istituto Nazionale di Alta Matematica, in Rome, at the beginning of February
1973. Among the invited speakers were well known computer scientists such
as John Reynolds, Maurice Nivat, Louis Nolin, Marcel-Paul Schützenberger,
Jean François Perrot, and also experts in information theory such as Bruno
Forte, professor at Waterloo. Unfortunately the majority of mathematicians
in Rome University at that time were not at all interested in theoretical
informatics and, although the program was of very high profile, beside our
group of IAC researchers only very few mathematicians attended the conference. One of them was Wolf Gross who had moved to the Department of
Mathematics and who was teaching a course of mathematical logic in which
computability issues had an important place. Beside giving my lectures at the
School on Automatic Control and Computing Systems, I was also collaborating with him. This gave me the opportunity to follow a few master's students.
In particular I advised the Master's theses of two brilliant students, Marco
Protasi[34] (on approximation notions in recursive functions theory) and Ma-
rina Moscarini (on subrecursive functions defined by recurrence relations).
Both later became professors at the University of Roma Tor Vergata and at
Sapienza University respectively and shared for a long way the trail of my
scientific interest.

I was also deeply involved in the activity of GRIT. Beside publishing the
bulletin we were engaged in the organization of the first Italian congress on
theoretical informatics. The congress was sponsored by GRIT, AICA and
the Pisa CNR institute IEI[35] and throughout the year 1972 several meetings

---

[34] Since then Marco was for me an excellent scientific partner and a dear friend until his
premature death in 1998, when he was only 48.

[35] The organizing committee consisted of the leaders of all main Italian groups working in
theory of computing: Corrado Böhm, Stefano Crespi Reghizzi, Gianni Degli Antoni, Antonio Grasselli, Fabrizio Luccio, Ugo Montanari, Marco Somalvico and I, plus Gianfranco

were held to define aim and scope of the conference and organization details. According to the call for papers the scope was focused on the following topics: theory of automata and formal languages, theory of algorithms and information structures, computability and complexity theory, artificial intelligence, theory and formal properties of programs, and semantics of languages and programs, but indeed the accepted papers expanded on a larger landscape of topics (parallel computation models, uniformly reflexive structures, λ-calculus, etc.). Concerning the aim, the conference was organized having in mind two objectives: presenting the best work that was currently being carried out by various Italian groups but also to provide to young researchers a view of the main directions that were being pursued at international level.

The congress was held in Pisa, in the historical building of the IEI institute, in Via Santa Maria, at the beginning of March 1973. We were all very excited. For the first time all Italian computer scientists who started to identify their own work with the new stream of theoretical research would gather together. The conference was organized over three days and the presentation of current hot topics took up the first two days and was achieved by means of tutorials presented by Italian researchers and invited talks given mainly by foreign scientists. So the first day Degli Antoni gave a tutorial talk on 'Properties of Programs' in which he illustrated various approaches to the analysis of properties and semantics of programs, Ugo Montanari gave a talk on 'Declarative and Procedural Epistemology' in which he discussed the problem of knowledge representation in artificial intelligence, Marco Somalvico spoke about 'Automatic Deduction', Giuseppe Iazeolla spoke about 'Parallel Computation Models' and, finally, I gave a talk on 'Analysis of Algorithms and Program Complexity'. The invited talks, given at the second day, addressed various issues that were considered hot topics in that moment. Corrado Böhm and Mariangiola Dezani (University of Turin) spoke about generating functions for representing families of data structures (extending previous work by Donald Knuth), Mike Paterson (University of Warwick) gave a talk on string matching with don't care symbols, Ashok Chandra and Zohar Manna[36] (Stanford University) spoke about the computational power of program schemes endowed with features such as counters, pushdown stacks, equality, recursion etc., Peter Lucas[37] (IBM Vienna Lab.) spoke about step

---

Capriz and Milvio Capovani (two renowned mathematicians who were respectively director and leader of the mathematics section of IEI) and Paolo Ercoli (representative of AICA).

[36] Zohar Manna, born in 1939, is a professor at Stanford University. Manna has made outstanding contributions to the formal study of computer programs and to the application of temporal logic in the analysis of concurrent and reactive systems. His book *The Mathematical Theory of Computation*, appeared in 1974, was one of the first textbooks addressing the properties of computer programs in mathematical terms.

[37] Peter Lucas, born in 1935 passed away in February 2015. He was one of the most important figures of the IBM Vienna Laboratory that in the 1960s and 1970s produced remarkable research in programming languages in general and more particularly in program development methodologies and program semantics. He was also among the first to introduce an axiomatic definition of data types.

by step program development based on Hoare's axiomatic approach, Thomas Bredt (Digital System Lab.) gave a talk on correctness analysis of concurrent programs, and Robin Milner[38] addressed a topic that was still rather new at that time but would explode in the 1980s: an approach to the analysis of properties of parallel processes in view of 'extending the present techniques for formal semantics to real parallel programming languages'. The contributed papers were on the last day. Ten papers had been selected out of 18 submissions. Most of the contributed papers could be essentially classified in the following domains: formal languages (formal power series, syntax directed translation, extensions of grammars to abstract structures), recursive program schemes and computation rules, λ-calculus and Scott's approach to semantics. It is now impressive to see that all young scientists who presented their papers at the conference became outstanding scientists in the subsequent decades: among them Beppe Attardi, Alberto Bertoni, Giorgio De Michelis, Carlo Ghezzi, Andrea Maggiolo, and Dino Mandrioli, although some of them later moved to more applied areas of computer science.

The conference was a very successful and lively event. I can remember heated discussions on the relevance of the various approaches to computational complexity, between Franco Preparata (Fig. 5.10) and me. Franco had studied in Rome at the Faculty of Engineering and had left in 1965 for the United States where he had joined the faculty of the University of Illinois at Urbana Champaign and had become professor in 1970. Later he had returned to Italy for a few years and in 1973, at the time of the conference, he was a professor in Pisa. Although his most known contributions came later with his outstanding work concerning interconnection networks, parallel computing and computational geometry, in 1973 he was already famous for his results regarding coding theory and fault diagnosis in circuits. His approach to algorithm design has always been very concrete and oriented toward practical solutions so he was rather skeptical about the abstract recursion theoretic ap-

Fig. 5.10 Franco Preparata

---

[38] Robin Milner (1934–2010) was one of the most influential British computer scientists. His name has already occurred in these notes due to his scientific role in the theory of programming and to his participation in the creation of EATCS. He was assigned the Turing Award in 1991 "for three major achievements: LCF, the mechanization of Scott's logic for computable functions, ML, the first language to include polymorphic type inference, and CCS, a general theory of concurrency".

proach to complexity. Also a great debate at the conference was stimulated by
the various concepts of program semantics that were presented in the talks.
The Pisa group was rather interested in denotational semantics and fixed
point theories while Degli Antoni's group in Milan was opposing the more
practical relevance of logic approaches in view of assessing the correctness of
programs.

Chapter 6
The Journal

Since the foundation of EATCS the creation of a scientific journal covering topics in theoretical informatics was considered a fundamental step, both for providing the association with an outlet for scientific results produced by its members and for promoting the new science in Europe.

As we have seen, at the beginning of the 1970s the spectrum of journals devoted to theoretical computer science was still rather limited. The main journals where European and US scientists used to publish their results were the *Journal of Computer and System Sciences*, *Information and Control* (now *Information and Computation*) and the *Journal of the ACM*. Also the *Communications of the ACM* at that time used to publish theoretical papers; for example the Böhm and Jacopini paper had been published in this journal. Other journals where theoretical papers could be published were: *Mathematical System Theory* (now *Theory of Computing Systems*), *Information Sciences*, the *Computer Journal* (the journal of the British Computer Society) and *Information Processing Letters*. Of course, frequently, theory papers were also hosted by journals devoted to mathematics and logic, such as the *Journal of Symbolic Logic*, the journals of the American Mathematical Society (where some of the fundamental Hartmanis papers had been published), the *Zeitschrift für Mathematische Logik und Grundlagen der Mathematik* or the Italian journal *Calcolo* (born in 1964 and extended to theoretical computer science in 1970).

The explosion of research work in theoretical computer science in the late 1960s and early 1970s generated the need to create more publication outlets for the results produced in this area. In 1971 the Society for Industrial and Applied Mathematics created a journal (*SIAM Journal on Computing*) devoted to 'mathematical and formal aspects of computer science and non-numerical computing'. The first issue was published in 1972 and immediately the journal reached an outstanding quality standard[1]. Again in 1971 another

---

[1] Among the editors of the first issue were Ron Graham, Juris Hartmanis, John Hopcroft, Richard Karp, Eugene Lawler and Richard Stearns, reflecting a strong interest in the direction of algorithms and complexity. Among the first papers published in the first volume

© Springer International Publishing AG, part of Springer Nature 2018   113
G. Ausiello, *The Making of a New Science*,
https://doi.org/10.1007/978-3-319-62680-2_6

journal started being published by Springer, *Acta Informatica*, still one of the major journals in the field. Also the French Association for Economical and Technical Cybernetics (AFCET), a scientific society that had been created in 1968 by merging various pre-existing associations and that used the term *cybernetic* as a general (and sufficiently vague) label for various scientific domains ranging from automation to operations research and informatics, decided to start a journal devoted to 'Informatique Théorique et Applications'. The new journal was a branch of RAIRO (the *Revue d'Automatique, Informatique et Recherche Operationelle* that had been founded in 1967) and the first issue was published in 1974. I was invited to be on the editorial board. In the first two issues there were, among others, papers by Louis Nolin, on universal algorithms, Bruno Courcelle, on deterministic grammars, Flajolet and Steyaert, on immune sets, Giuseppe Longo and Marisa Venturini Zilli, on the complexity of theorem proving and Mariangiola Dezani Ciancaglini, on string manipulation.

The need to create a new journal was in any case discussed extensively in EATCS and references to this need were frequently expressed following the first meetings and in the letters that were exchanged among the founders.

The first EATCS General Assembly was held 24–26 March 1973 in Warwick. Beside the general assembly the program included a technical part consisting of a few scientific talks. The invitation from Mike Paterson[2] (Fig. 6.1) (vice-president of the association) was written in Mike's typical witty style:

Fig. 6.1 Mike Paterson

---

were a few historical papers such as *Depth-First Search and Linear Graph Algorithms* by Tarjan (that by now counts about 5000 citations), *The Transitive Reduction of a Directed Graph* by Aho, Garey and Ullman and *Some Undecidability Results for Parallel Program Schemata* by Miller.

[2] Michael Stewart (Mike) Paterson has been a professor at the University of Warwick since 1971. He obtained his PhD in Cambridge under the advisorship of David Park. He is one of the greatest scientific leaders in the field of analysis and design of algorithms and computational complexity. Among his students are outstanding scientists such as Leslie Valiant, William McColl and Costas Iliopoulos.

"If you would like to talk (in any language you like), please send me a title and a time estimate. I can guarantee you a room and a blackboard but not, of course, an audience." And also: "If you let us know your time of arrival in advance we can meet you at either Birmingham Airport or Coventry Station. Wave a magnetic tape or carry a Turing machine for identification."

At the general assembly (see Appendix I) the discussion concerned the aims of the association and the actions to be carried out to further such aims. In particular we decided to circulate an informal news bulletin, to organize a regular formal conference (like the conference held at INRIA the year before) and to support a series of specialized working conferences. The general assembly also elected the council (Ausiello, Böhm, Brauer, de Bakker, Mayoh, Milner, Nivat, Paterson, Perrot, Scott, Sintzoff and Verbeek; three council places were left vacant for the future invitation of colleagues from Israel and other countries). Then the next day the council met and elected the executive officers: Nivat, president, de Bakker and Paterson, vice-presidents, Sintzoff, treasurer and Mayoh, secretary. The issue of the creation of a journal was not addressed at the time, but it would be considered just three months later.

On 18 June 1973, Maurice Nivat wrote a letter to the council (see Appendix J) in which he said: "I send you a copy of a proposal from North Holland Publishing Company regarding the creation of a new journal of Theoretical Computer Science. This proposal was made after discussions I had with Mr. E. Fredriksson in which we asked ourselves whether it would be possible to have our European Association play a role in this creation. The idea we came to was that the association could appoint say five members of the future editorial board among whom, hopefully, one would be the managing editor and that these five people would discuss with North Holland the numerous questions to be solved: other members of the editorial board, scope of the journal etc. An interesting idea was to make the journal easily available in eastern European countries by having it also sponsored by some official organism from one of these countries". Also in this case it is worth reading what Maurice remembers about the creation of the journal *Theoretical Computer Science* (TCS): "The idea was not mine nor that of Schützenberger: the idea came from Einar Fredriksson who was editor in the Dutch publishing company North Holland; he came to see me and suggested that I create a journal, I do not know, and shall never know, how this idea came to his mind."[3]

Nivat's plan was to discuss the issue in the council meeting that was scheduled for October 1973 in Hamburg, hosted by Wilfried Brauer[4]. The minutes

---

[3] Maurice Nivat, *The True Story of TCS*, Virtual Special Issue for TCS 40th Anniversary, *Theoretical Computer Science* (http://www.journals.elsevier.com/theoretical-computer-science/virtual-special-issues/40th-anniversary-of-theoretical-computer-science/)

[4] Wilfried Brauer (1937–2014) was one of the fathers of German computer science. First a professor in Hamburg and then in Munich, he devoted a lot of energy to the development of computer societies. From 1994 to 1997 he was president of EATCS, then he was president

of this meeting, signed by Brian Mayoh, show that the engine of EATCS started to warm up. First of all the three remaining positions on the council were filled by inviting Erwin Engeler (Switzerland), Zohar Manna (Israel) and Arto Salomaa (Finland). It was also decided that Maurice Nivat was in charge of editing the first issue of the bulletin and that the bulletin would be printed by IRIA and distributed to all members of EATCS and also to some selected 'potential' members. Concerning membership, the approach toward non-European members was, in those early times, rather shy: "It was also agreed that the association could accept applications from non-Europeans but that it should not campaign for such members"[5]. A strong effort would instead be made to invite members from eastern Europe and to this aim it was decided to prepare a leaflet describing the association, to be sent to academies in eastern European countries "in an attempt to elicit their cooperation in our effort". Another decision concerned the sponsorship of a few European events. First of all the ICALP conference that at that time was still supposed to take place every two years: the second ICALP to be held in Saarbrücken in 1974 and the third to be held in Warwick in 1976 (but indeed the third ICALP took place in Edinburgh). Then the Italian congress on λ-calculus that Corrado Böhm was organizing in Rome in 1975, a Franco-Polish conference to be held in Warsaw and an Advanced Course on Foundations of Computer Science to be held in Amsterdam in 1974. Finally, at the council meeting Maurice Nivat discussed his contact with the publisher North Holland regarding "a new journal in theoretical computer science".

The relationship between EATCS and TCS has always been rather controversial. Although the original idea of Maurice Nivat was to involve EATCS directly in the creation of the journal[6], after his report on his contact with North Holland at the EATCS council in Hamburg the minutes of the meeting say: "This question does no longer concern the Association". This separation is confirmed in a letter sent by Maurice Nivat to EATCS council members at the end of 1973 in which the next meeting of the EATCS council is called for 23 February 1974 in Saarbrücken. In a postscript to the letter Maurice says: "Though it no longer interests directly EATCS, I'll give in Saarbrücken information about the journal *Theoretical Computer Science* created by North Holland. Let me say now that the negotiations progressed favourably and the creation of the journal will be announced very likely in January for a first issue to appear in October." Also in a letter sent to all EATCS members in

---

of the German GI until 2001. Wilfried and his wife Ute were lovers of Italian opera and never missed the most important performances.

[5] Actually quite rapidly the number of non-European members increased and EATCS became, in a few years, a truly worldwide organization.

[6] "When I decided to try to do so [to found the journal], in a natural way I asked to become editors of TCS the people who had been active in creating the association EATCS, the journal would be called TCS and a subtitle would indicate that it was the official journal of EATCS". Maurice Nivat, *The True Story of TCS*, Virtual Special Issue for TCS 40th Anniversary, cit.

1976 Maurice Nivat said: "Among our first aims was the creation of a journal: for various reasons the association is not directly linked with *Theoretical Computer Science*, the new journal published by North Holland. Nevertheless it inspired it and seven members of the council of EATCS stand on the editorial board of TCS."

Unexpectedly, in a subsequent letter sent in the same year, Nivat suggested that through an agreement between EATCS and North Holland the subscription rate of TCS for all EATCS members might be fixed at 10 USD. Apparently the discussion between Nivat and North Holland concerning the relationship of TCS and EATCS was still progressing, and finally the situation changed again in June 1978. In issue N. 5 of the bulletin Nivat provided a long report on an agreement between EATCS and TCS that starts with the words: "The journal *Theoretical Computer Science* will be published from now on as 'The journal of the European Association for Theoretical Computer Science'. Precisely this is what will be printed on the front cover of TCS below the title which remains Theoretical Computer Science. This change results from an agreement which has been recently signed by a representative of North Holland, the President of EATCS (M. Paterson) and the editor-in-chief of TCS (M. Nivat)". After explaining that there was no interference between the journal TCS and the bulletin of EATCS, Nivat's report goes on with the following words: "This means that the editor-in-chief of TCS takes the engagement (even if this is not written in the above mentioned agreement) to keep the EATCS informed of every change in the editorial board or in the editorial policy, to send his resignation to the president of EATCS at the same time as to North Holland Publishing, to resign immediately if the general assembly of EATCS emits a note towards this effect. He also takes the engagement to give periodically information about the status of TCS in this bulletin, including titles of forthcoming papers, length of the backlog and whatever seems proper to be brought to the knowledge of EATCS members". The agreement was ratified by the EATCS general assembly in July 1978 (see bulletin N. 6) and was in effect throughout the years 1980–1991. From volume 12 to volume 80 the journal carried the words 'The journal of the EATCS' on the front cover and quite often Maurice reported the relevant changes in TCS policy to the general assembly (for example when the split between track A and track B was decided in 1991).

At the end of the 1990s this special relationship came to an end without a clear explanation. Presumably EATCS was interested in adopting a more open attitude toward not just one, but all journals devoted to theoretical computer science. Besides some concern about the pricing policy adopted by the publisher, that in the meantime had changed from North Holland to Elsevier, had encouraged the association to loosen the ties with TCS. Despite this change in the special relationship between the association and the journal, it is in any case evident that, beyond any doubt, the creation of the journal was one of the most important consequences of the birth of EATCS and of the role that Maurice Nivat played in both events.

Maurice Nivat was president of EATCS from 1973 to 1977 and, beside the foundation of the journal *Theoretical Computer Science*, during his term of office the other two major achievements of EATCS materialized: the establishment of the *Colloquium on Automata, Languages and Programming*, which after the first edition held in Paris in 1972 was held in Saarbrücken in 1974 and in Edinburgh in 1976, becoming an annual event in 1977 when the conference was held in Turku, and the creation of the EATCS bulletin.

The first issue of the bulletin was published and distributed in December 1973. The editorial written by the president provides a picture of what EATCS was at that time (about fifty members, running the ICALP conference every two years, sponsoring a few scientific events, establishing agreements and exchanges with SIGACT, the American sister organization) but also what theoretical computer science itself was. The editorial says (see Appendix K): "A general agreement was reached to consider as the major points of theoretical computer science the following fields of interest: (i) Theory of automata (ii) Formal languages (iii) Algorithms and their complexity (iv) Theory of programming, if such a thing exists: we may now consider as chapters of this theory formal semantics, proving properties of programs, and all applications of logical concepts and methods to the study and design of programming languages. This list is by no means limitative: defining the limits of theoretical computer science is at least as difficult as defining the limits of computer science itself. And we strongly believe that a science is what the scientists at work make it: certainly new areas of computer science will be open to theory in the near future". Other interesting information contained in the bulletin was a report on the MFCS conference in Slovakia and surveys of activity in various research centers.

The symposium and summer school on Mathematical Foundations of Computer Science, the second in the series, had taken place at the beginning of September 1973 in Strbske Pleso (in Slovakia, in the High Tatras), and had been organized by the Slovak Academy of Sciences and by the United Nations Computing Research Centre in Bratislava. The program committee was chaired by Jiří Becvar[7] and consisted of Józef Gruska, Ivan Havel, Jiří Horejs and Branislav Rovan. Arto Salomaa reports in the EATCS bulletin that the conference had been attended by 140 participants and that 50 lectures had been presented. I was at the conference and that was my first trip beyond the 'Iron Curtain'. Beside the scientific atmosphere of the conference I remember the enchanting beauty of the Tatra Mountains and the gipsy music played in restaurants, but also the mixed feelings that struck me when I arrived in Prague. The red flags waving along the freeway from the airport to the city and the Soviet style slogans written on the public buildings were creating conflicting sensations in a country that only five years before had seen a dramatic repression following the 'Prague Spring'. At the weekend I took the opportunity to visit Prague's cemetery, looking for Jan Palach's grave.

---

[7] Jiří Becvar (1926–1999), a logician, promoted the development of theoretical computer science in Czechoslovakia.

In any case, the program of the conference was very interesting. The formula that has characterized MFCS since the beginning, that is a clever combination of invited papers and contributed papers, made the conference very fruitful for the participants. At the second MFCS there were 18 invited talks and 32 contributed talks. Four lectures were in Russian. The main research directions that were presented in the invited lectures were primarily formal languages and automata theory (Geczeg, Ginsburg, Gruska, Nivat, Salomaa etc.), followed by complexity (Hartmanis and Schnorr). The Czech school of logic was represented by P. Hajek who presented his approach to inductive information extraction from experimental data (today we would classify his contribution as 'data mining'), while the Polish school of semantics was represented by Andrzej Blikle who gave a talk on an algebraic approach to programs and computations. Finally a survey on Petri nets was given by Petri[8] himself in one of his rare public appearances.

Beside the report on MFCS, as I said, the first issue of the bulletin also contained surveys of research activity. A view of these reports is interesting for understanding the European scene in 1973. One report, prepared by Jaco de Bakker, concerned the activity in The Netherlands. At the Mathematisch Centrum Peter van Emde Boas was working on resource bounded complexity classes, while de Bakker and de Roever were working on properties of programs and of program schemes, and Paul Vitanyi's research was devoted to cellular automata (in particular 'sexually' reproducing automata) and to the study of Lindenmayer systems as parallel rewriting systems. At TU Delft the main work was carried out by Fokkinga (semantics of programs), Verbeek (formal languages) and van der Poel (combinatory logic and λ-calculus). At TU Twente the activity in theoretical computer science was quite recent and involved only one assistant professor! Utrecht and Leiden were not even mentioned.

Another report concerned Turin University. λ-calculus was of course the most relevant research topic: Corrado Böhm, Mariangiola Dezani, Simona Ronchi Della Rocca and Carla Simone were studying different evaluation rules for λ-expressions and were also starting to address typed λ-calculus, a hot topic to which Turin researchers would make important contributions. Other topics addressed in Turin were families of information structures and their generating functions, and the relationship between parallel computation and the Church-Rosser property. Finally, a report concerned the 'Institut de Programmation' at Paris VI. In this institute, which strictly collaborated with Nivat and Schützenberger at the University of Paris VII, the major topics were automata and formal languages (Max Fontet, Jean François Perrot, Jacques Sakarovitch and Jean Berstel), complexity of algorithms (in particular complexity of decidable properties in the formal theory of programs:

---

[8] Petri invented the nets that took his name in 1962 in his PhD Thesis, at the University of Darmstadt, whose title was *Communication with Automata*. Petri nets became very popular in the 1980s as a model for describing the workflow in office and industrial automation. Petri died in 2010 at the age of 84.

Bénéjam and Vidal-Naquet), formal definition of data structures (Bernard Robinet) and program schemes (G. Cousineau and J-M. Rifflet).

In December 1973 I wrote a report for the director of IAC, Ilio Galligani, and for the Mathematics Committee of CNR concerning the developments of EATCS and of the various initiatives that were materializing. This was also my farewell to IAC.

The end of 1973 was going to bring an important change that would affect the rest of my professional life. For a few months Paolo Ercoli had been suggesting that I should move from IAC to CSCCA, the CNR research center founded by Antonio Ruberti at the Faculty of Engineering of Rome University. Two researchers affiliated with CSCCA, Giuseppe Iazeolla and Daniel Bovet, had moved to Pisa, another researcher had left CNR to take a position in industry and the perspective was that in one or two years Paolo would become full professor at the Faculty of Engineering. The informatics group of CSCCA would need to be completely rebuilt and this was the task that Ruberti and Ercoli were offering to me. The decision was not easy for me. On one side I was tired of working in a mathematics environment that had an instrumental view of computers (the new buzzword that they were using to identify their scientific interest was 'computational mathematics' but they were not showing any interest for 'computational' aspects), and that did not even consider theoretical computer science a real science. On the other side the risk was falling into the arms of another strong group (Ruberti's group in control and system theory) that might just aim at extending their academic power to computer science. I finally decided to trust Ruberti. One of the positive aspects that encouraged me to accept the invitation was that Ruberti and his young colleagues, most of them his former students, were known for the high profile of their research work (Alberto Isidori, in particular, who would become one of the world's greatest scientists in the field) and this was a big incentive to join. Besides, although authoritative in his own field and strongly persuaded of the role that system theory could play in most human fields, from technology to economics, Ruberti was a smart and alert observer of the changes in science and society, and he had started to realize the increasing role of informatics in industry, management and public administration. So he was apparently sincere when he said he wanted to expand informatics research in his center.

On 1 January 1974 I changed my affiliation. As leader of a small but potentially strong group that already included three young bright researchers, Carlo Batini, Alberto Pettorossi and Silvio Salza (all of them now professors at various Italian universities), plus Marina Moscarini who had just received a scholarship from CNR, I had to establish the new research directions of the informatics section of the center, melding my personal scientific interest in algorithms and complexity, a subject that Ruberti (with his farseeing vision) considered important and fascinating, with some of the major and trendy research topics that could allow our young people to find their own way in the scientific community. Structured programming, properties of programs and data structures and information management systems were some of the subjects that we decided to address at the beginning, with the aim of showing that the new group could combine theoretical research with solid and practical informatics issues. Besides the team would also address topics in the field of computer systems (such as, for example, performance evaluation) under the supervision of Giacomo Cioffi who was still a researcher at Fondazione Ugo Bordoni. For me it was an important achievement. The great development that was taking place, which I really perceived as an extraordinary step forward in that moment, was that for the first time in Rome a research center would include relevant objectives in theoretical informatics in its annual research program.

My new affiliation had various other positive implications. First of all I was expected to be officially in charge of a course in the electronics curriculum of the engineering faculty: Compilers and Operating Systems. Although the title seemed rather weird (compiler construction and operating system design relied on quite different theoretical and pragmatical principles, formal languages and syntax analysis in the first case and, in the other case, algorithms for concurrent processes and resource management) it was still a good chance to provide engineering students with some elements of the foundations of computer science. Moreover, the support of the director of CSCCA, Antonio Ruberti, allowed me to plan some long visits abroad. I had two standing invitations, one by Maurice Nivat to spend one month in Paris and another by Juris Hartmanis to spend four months at Cornell University. So I decided to go to Paris in May 1974 and to postpone the visit to US until 1975.

My stay in Paris was very exciting. The way in which Philippe Flajolet (Fig. 6.2) and Jean Marc Steyaert received me was gorgeous and started with a superb dinner at Les Halles, the commercial neighborhood where, in that period, a huge hole stood in the ground, similar to a gigantic quarry, the place where the 'new Halles' would be built[9]. I was supposed to give some lectures on computational complexity to young researchers and graduate students based on the book that I had just finished *Complessità di calcolo delle*

Fig. 6.2 Philippe Flajolet (left) with Wolfgang Panny

---

[9] The year before, the hole had been chosen as the set for a western movie shot by the Italian director Marco Ferreri, *Non toccare la donna bianca*, in which the story of the defeat of General Custer by Chief Sitting Bull in the battle of Little Big Horn was revisited in modern terms.

funzioni (subrecursive classes, Turing machine complexity, abstract computational complexity, basic principles of design and analysis of algorithms). I
was also supposed to exchange ideas and do some research work with Philippe
and Jean-Marc.

My office was at Paris VII in the Jussieu buildings but once or twice a
week I went to IRIA with my hosts, to visit and talk to the fantastic people
that Maurice Nivat had gathered there: Gilles Kahn (who was already well
known for having invented a formalism, the 'Kahn networks', for describing
distributed computations and who later became one of the most important
world experts in semantics of concurrent programs[10]), Jean Jacques Levy
(engaged in his work on λ-calculus), Gérard Huet (at that time working on
unification in typed λ-calculus and on the project of the mentor program
editor that he was carrying out with Gilles Kahn), Jean Vuillemin (who was
moving from his work with Manna on operational semantics of recursive programs to the area of design and analysis of algorithms) and many others. Of
course Philippe Flajolet and Jean-Marc Steyaert also had an important role
in the theory group at IRIA. Their work, formerly oriented toward recursion
theory and abstract computational complexity, was already switching toward
more practical issues regarding the average performance of algorithms and
data structures.

One of the memorable events that took place while I was staying in Paris
was the defense of the 'Thèse d'état' of Bernard Robinet[11]. For his research
work and for the promotion of scientific events Bernard deserves a few words.
In 1974 Bernard was one of the most brilliant researchers at the 'Institut de
Programmation' in Paris VI. He was doing research on λ-calculus and on
formal semantics of programs. He was also studying the notion of completeness for programming languages and devoted quite a lot of time to the study
of APL. In April 1974 he had organized in Paris the first *Colloque sur la
Programmation*, the first conference aiming at putting together the theory
and practice of software. The conference had been very successful. All topics
that were currently hot were addressed: structured programming, program

---

[10] Gilles Kahn became scientific director and then president of INRIA from 2004 until his
premature death in 2006. Let's see how his colleagues Philippe Flajolet and Gérard Huet re-
call his scientific work: "Gilles Kahn a été l'un des précurseurs du domaine de la sémantique
des langages de programmation, laquelle permet de donner un sens mathématique précis
à un programme informatique. Il contribue notamment, en collaboration avec l'Université
d'Edimbourg où il passera une année sabbatique en 1975–76, à l'élaboration d'une importante théorie des domaines de calcul concrets. En 1983, il participe à la création de
l'Unité de recherche INRIA de Sophia-Antipolis. Il y développe une théorie élégante, celle
de la "sémantique naturelle". On peut grâce à cela manipuler un programme comme une
formule, c'est-à-dire calculer des propriétés de programmes. Cette percée ouvre la voie au
développement d'environnements de programmation, où les programmes sont manipulés
de pair avec leur spécification".

[11] Bernard Robinet died in 2009, at the age of 68. After a long career in the University of
Paris VI, where he did research on λ-calculus and formal semantics of programs and where
he also became vice-president, he moved in 1992 to Télécom ParisTech as director of the
education programs and of the Doctoral School.

correctness, program properties, program synthesis, control structures, program schemes, parallel processing, data flow, formal semantics and theory of types, and all major figures in the theory of programming were present (among others Louis Nolin, Adriaan van Wijngaarden, Peter Naur, Ole Johan Dahl, John Reynolds, Butler Lampson, Jack Dennis and the Italians: Corrado Böhm, Gianni Degli Antoni, Luigia Aiello and her husband Mario, and Renzo Pinzani).

The creature of Bernard Robinet, the *Colloque sur la Programmation* (aka the *International Programming Symposium*) went on for 12 years until in 1986 it was merged with the *GI Workshop on Programmiersprachen und Programmentwicklung* (Programming languages and program development) to give birth to the *European Symposium on Programming* (ESOP). The chairs of the first ESOP were Bernard Robinet and Reinhard Wilhelm. ESOP still is the main European event on the theory of programming and is part of ETAPS, the *European Joint Conference on Theory and Practice of Software*.

But Bernard Robinet was also enjoying life. The 'pot de thèse' that followed the defense was gigantic, worthy of Rabelais' Pantagruel. The table was covered by a huge amount of 'foie gras', by all possible kinds of cheese and by an uncountable number of wine bottles coming from all regions of France.

My stay in Paris was also a great chance to establish a warm friendship with the lovely family of Maurice Nivat, his wise and sensible wife Paule, a professor of mathematics, the pillar that supported the life of the whole family, and their children, Dominique, Jean-Luc and Catherine. I also had the chance to get acquainted with the city and in particular with the Latin Quarter where I was living. The atmosphere in Paris, in the wake of the first elections in which François Mitterrand and the French Socialist Party might have gained power (which indeed did not happen!), was full of excitement. I remember that the chances of Mitterrand winning the elections was the major topic of discussion during the trip I made with Nivat's family to Latrecey, to see the beautiful country house that Maurice and his wife Paule had just bought. The entire city, the entire country, were excited and touched by a leftist mood. Once I was shocked to see that in a movie theatre showing the Italian movie *La Battaglia d'Algeri*, by Gillo Pontecorvo, concerning the liberation fight of the Algerian people, leftist students were applauding terrorist attacks against the French Army in the Algiers Kasbah. I also felt 'gauchiste' shivers when I had dinner in a small Vietnamese restaurant close to Jussieu, run by an old Viet Cong militant, with Ho Chi Minh pictures on the walls.

One of the most important events that was held in that month, while I was staying in France, was the *Spring School on Complexity of Algorithms* organized by Philippe Flajolet on the Ile de Berder, off the coast of Brittany. The *Ecoles de Printemps d'Informatique Théorique* were another creature of Maurice Nivat and after more than forty years they are still alive and well[12]. This was the second spring school, the first one had been organized in 1973 by Jean Paul Crestin and Maurice Nivat at Bonsacre on the subject of 'Langages algébriques' (context-free languages).

The location was very beautiful and the visit to the fascinating and mysterious menhir alignments at Carnac still is, for me, a wonderful memory of that week. The program of the spring school was great. Beside a few more or less classical talks, including my own, the real great attraction of the school was a young brilliant scholar, who was about to get his PhD from Stanford: Ron Rivest. At that time Ron Rivest was already famous for having contributed, together with Manuel Blum, Robert Floyd, Vaughan Pratt and Robert Tarjan, to the design of linear algorithms for selecting the i-th element and computing the median of a set of numbers, and for his work on algorithms for partial match retrieval, but his major contribution to computer science would arrive just a few years later[13].

Almost at the same time another event took place in Amsterdam, the Advanced Course on the Foundations of Computer Science, where Jaco de Bakker lectured on programming theory, Mike Paterson about matrix algorithms and Erwin Engeler about computability. But an interesting fact about that school was that new topics also started to be addressed: theoretical aspects of operating systems were illustrated by Nico Habermann[14], Erich Neuhold lectured about the formal foundations and properties of databases

Fig. 6.3 From left to right: Adi Shamir, Ron Rivest and Leonard Adleman at the time of their joint work on cryptography

---

[12] The recent *Ecole de Printemps en Informatique Théorique* in 2016 was devoted to: 'Graphes et surfaces: algorithmique, combinatoire et topologie' and was held in Marseille.

[13] The work that was later done by Ron Rivest had a terrific impact on cryptography and network security. Only three years after the school on the Ile de Berder, when he was thirty years old, Ron Rivest, together with Adi Shamir and Leonard Adleman, invented the famous RSA (from the initials of their names) public-key cryptography scheme, just a few months after that Diffie and Hellman published the first cryptographic system based on asymmetric keys. For the relevance of their work in cryptography Adleman, Rivest and Shamir obtained the Turing Award in 2002 (Fig. 6.3).

[14] Nico Habermann, one of the fathers of Dutch computer science, died in 1993 at the age of 61. He was known for his research work on interprocess communication and for having cooperated since the beginning in the development of early operating systems (e.g. the THE operating system designed by Edsger Dijkstra).

(a research domain that would explode a few years later) and Bob Kowalski presented the foundation of the programming language PROLOG in his lectures on predicate logic as a programming language. This talk anticipated the presentation, with the same title, that Kowalski made at the 1974 IFIP World Congress the next summer and that (together with Rabin's talk on the hardness of logical decidable theories) would be another historical landmark in the development of the new science.

After my stay in Paris I went back to Rome. My suitcase was full of records of Georges Brassens. But soon it was time to leave again. At the end of July, in Saarbrücken, the second ICALP was being held. I went there by train with my three year old son Gabriele, having planned to proceed by train from Saarbrücken to Stockholm, where the IFIP Congress *Information Processing 74* was going to take place on 5–10 August. The *Second International Colloquium on Automata, Languages and Programming* was organized by Jacques Loeckx at the University of Saarlandes (the Max Planck Institute for Informatics would be created only fifteen years later). The program committee consisted of C. Böhm, W. Brauer, E. Engeler, P.C. Fischer, S. Ginsburg, J. Hartmanis, G. Hotz, J. Loeckx, M. Nivat, D. Park, Z. Pawlak, M. Rabin, A. Salomaa, M.-P. Schützenberger and A. van Wijngaarden.

The major role in the program was played by formal languages (in particular Adrian van Wijngaarden gave a talk on two level grammars and Arto Salomaa spoke about parallelism in grammars as it was provided by Lindenmayer systems), but semantics (e.g. de Roever and Reynolds) and properties of programs and program schemes (e.g. Chandra, Martelli and Montanari) were getting more room. The field of algorithms and complexity was dominated by complexity papers (Monien, Hartmanis, Mehlhorn and Weihrauch) and very few algorithmic contributions (two papers by Patrick Fisher on matrix algorithms) were presented. The presence of eastern European countries was very limited. Gecseg (from Szeged) spoke about the composition of automata and Witold Lipski[15] (from Warsaw, only 25 years old, in one of his first presentations abroad) spoke about file organization. Finally Antoni Kreczmar (also from Warsaw) presented the Polish version of logic programming, that is a language based on algorithmic logic (LOGLAN), designed to incorporate in a program the proof of its correctness.

Again in this edition of ICALP talks were given not only in English but also in French and in German. The proceedings appeared as volume 14 of the newly born Springer series of *Lecture Notes in Computer Science* (LNCS). All ICALP proceedings were published since then, until 2015, in the Springer series LNCS, with the only exception the third ICALP held in Edinburgh in 1976.

---

[15] Witold Lipski had been educated in the school of information retrieval created by Pawlak in Warsaw. In the subsequent years he made excellent contributions in the field of physical file organization (in particular related to the consecutive retrieval property). Witold Lipski died in 1985 when he was only 36.

The series LNCS is another success story in the field of theoretical computer science and its life is strongly tied to the history of ICALP and of the other major conferences in this domain. The series had been conceived in 1972 by Gerhard Goos, a professor at Karlsruhe, and by Juris Hartmanis, while he was on sabbatical in Germany. Previously, topics concerning theory of computing used to be published by Springer in the series of *Lecture Notes in Mathematics*. It was an important step by Goos and Hartmanis to make the proposal, but it was also a great intuition of the publisher to foresee that computer science would generate such an explosion of conferences and related proceedings. For a long time the editors of the series were Goos and Hartmanis, several years later joined by Jan van Leeuwen[16] and by a large editorial board. The first four volumes of the series appeared in 1973. Volumes 1 and 2 were devoted to conferences of the *Gesellschaft für Informatik* (GI). The German informatics society had been founded in the 1960s and its first annual congress took place in 1971. Among the first interest groups that were created was the group of automata theory and formal languages. Volume 1 of the LNCS series contained the proceedings of the 3rd congress of GI, edited by Wilfried Brauer, that included papers on automata theory, efficient algorithms, and the syntax and semantics of programming languages, plus various papers on applied topics. Volume 2 contained instead the proceedings of the first *Fachtagung über Automatentheorie und Formale Sprachen*, that is the first meeting of the group on automata theory and formal languages. The meeting had taken place in Bonn in July 1973 and had four top invited speakers: Erwin Engeler, who spoke about the structure of algorithmic problems, Zdzislaw Pawlak, who lectured about computing models, Michael Rabin, who spoke about logical theories related to formal languages and finally Arto Salomaa, who spoke about developmental languages based on Lindenmayer systems. Klaus Indermark and Karl-Heinz Böhling were the organizers and the editors of the proceedings. It is interesting to know that volumes 3 and 4 of LNCS were devoted to the 5th *Conference on Optimization*, an IFIP event that had taken place in Rome in 1973, rather than to a computer science event. The editors of the proceedings were Roberto Conti and Antonio Ruberti!

As planned, after ICALP I continued my trip by train to Stockholm. The journey turned into a real nightmare because my son was very excited and did not sleep for 24 hours, but the beauty of the city of Stockholm was good compensation for the stress of the trip. Also the quality of the conference was rewarding. Throughout the 1960s and the 1970s IFIP maintained its tradition of being the world's main computer organization and its conferences were attended by thousands of computer scientists working both in applied domains (the conference also hosted an industrial exhibition) and on

---

[16] Jan van Leeuwen (born in 1946) is one of the most productive and highly cited leaders of European computer science. His scientific interests span formal languages, analysis and design of algorithms, online algorithms, distributed algorithms and computational geometry. He has been a professor at Utrecht since 1977.

theoretical issues. The preceding conferences in 1968 in Edinburgh and in 1971 in Ljubljana had been very successful and, judging from the program, the conference *Information Processing* 74 seemed even more promising. The conference was organized in six tracks: computer hardware and architecture, software, mathematical aspects of information processing, technological and scientific applications, applications in the social sciences and the humanities, systems for management and administration.

Although an explicit title referring to theoretical informatics was not present[17], the track regarding software and the track regarding mathematical aspects of information processing (in which papers on genuine theory of computing subjects were mixed with papers regarding numerical analysis, differential equations, nonlinear optimization etc.), were rich in relevant theoretical content, and some of the talks had a great historical value.

In the software track two invited papers had an important role in future developments. One is Rod Burstall's paper on *Program Proving as Hand Simulation with a Little Induction* in which (quoting Burstall) "Manna's method is stripped of its predicate calculus formalism ... and dressed in new clothes borrowed from Deutsch[18]". In other words, the analysis of program correctness is performed by making use of symbolic execution and a less formal, but more intuitive, use of induction. The second is Mealy's paper which contains the first debate about different approaches to the formalization of data structures (axiomatic versus operational).

In the mathematical aspects track two papers were presented that have been already mentioned in preceding chapters as having a big impact in the field: one is Kowalski's paper on *Predicate Logic as Programming Language* (one of the theoretical pillars on which logic programming stands, a paper that has by now gained almost 1300 citations) and the second is Rabin's invited talk on *Theoretical Impediments to Artificial Intelligence*, in which the inherent complexity of apparently simple decidable theories is analyzed. At the time this paper created a lot of discussion because it contradicted the claim, often supported by researchers in AI, that stating a problem in logical terms makes the solution easier. On the contrary, embedding a problem in a logical theory (even though decidable), can make the solution of the problem utterly unfeasible. Concerning the impact of his talk, it is interesting to see what Michael Rabin (Fig. 6.4) himself recalls: "I described results like the intractability of Presburger arithmetic – some AI people were writing theorem proving programs and here was a result, in a very simple case, proving

---

[17] In 1977, when the IFIP World Congress was held in Toronto, a theory conference was organized in Waterloo by E.A. Ashcroft, J.A. Brzozowski and K. Culik II back to back with the IFIP event. Formal theory tracks at IFIP congresses started to appear explicitly only after the creation of the Specialist Group on Foundations of Computer Science in 1990 and the subsequent creation in 1996 of Technical Committee 1: Foundations of Computer Science.

[18] Peter L. Deutsch, whose PhD thesis at Berkeley concerned an interactive program verifier based on partial symbolic execution.

Fig. 6.4 Michael O. Rabin

that this is absolutely hopeless. The theorem proving people were extremely worried. They were afraid – they told me so – of the impact this would have on the funding agencies."[19]

Beside these papers it is worth remembering two talks concerning the relational model of databases: Edgar Codd's talk *Recent Investigations into Relational Data Base Systems* and William Armstrong's talk *Dependency Structures of Data Base Relationships*, a subject that prompted a large number of contributions in the subsequent development of database theory. This paper now has more than 1000 citations. It is also interesting that for the first time parallel computing gained meaningful attention and a series of papers (by Ashok Chandra, Gilles Kahn, Thomas Bredt and Joe Traub) on parallel computation models and parallel algorithms appeared in the track.

On the way back from Stockholm I stopped in Dogliani, my home village, in the Langhe district, where I had my vacation and where Philippe Flajolet came to visit me. We had a few days of pleasant and relaxing vacation, good food and excellent wine, with just enough scientific work to make it even more fulfilling. My book on computational complexity was about to be printed by the Italian publisher and Philippe agreed to make an English version.

In the meantime plans for the future piled up. I had the invitation by Juris Hartmanis to spend four months at Cornell during the summer of 1975 and the trip still had to be organized. In the spring of 1975 Corrado Böhm was having a conference on λ-calculus in Rome and I was involved in the preparation. Moreover in November 1974 GRIT was going to celebrate the second congress on Theoretical Computer Science. And also, in November I was supposed to start my new course on compilers and operating systems for

---

[19] D. Shasha, C. Lazere, *Michael O. Rabin. The possibilities of chance*, in *Out of Their Minds. The Lives and Discoveries of 15 Great Computer Scientists*, Copernicus, 1995

fifth year students in electronic engineering. Finally, very soon in October my third child, Sara, was expected!

The *Convegno di Informatica Teorica* was held in Mantova on 21–23 November 1974. One of the invited speakers was Robert Constable and he gave a very beautiful survey talk on the complexity of decidability of equivalence of flowcharts and of recursive programs defined on various algebraic structures (integers, lists etc.). Beside invited talks the program contained ten contributed talks by some of the most active research groups: Turin, Milan, Pisa, Naples and Rome. In particular Böhm, Dezani and Ronchi della Rocca spoke about encodings of sets and multisets, Bertoni addressed counting problems in weighted context free grammars, Luccio, Lodi, Pagli and Santoro spoke about partitioning and search problems on information structures, Martelli and Montanari presented their fix point approach to the characterization of the solution of dynamic programming algorithms, Venturini presented a paper on unification and Moscarini presented the work we had done together on subrecursive classes based on recurrence relations.

My course at the University of Rome had me involved almost full time for the rest of the academic year 1974–1975. During that year the students following the curriculum of computer engineering had three courses in which computer science content was given, not much but enough, at that time, to give them a good computer science education: the electronic computer course that Paolo Ercoli had been teaching for almost a decade, the switching circuit course of Giacomo Cioffi and my course on compilers and operating systems; there was a great expectation by the students. When I entered the classroom for my first lecture it was full with about one hundred electronic engineering students. I was also impatient to teach all the fascinating topics that I had happened to learn and that appeared to me the inevitable scientific background of anybody (scientist or professional) dealing with computers. Starting with formal languages and automata theory, and going through computability based on Turing machines, I decided to address also some fundamental algorithm design techniques and information structures (inspired by Fabrizio Luccio's book *Strutture, linguaggi, sintassi*) and a little bit of computational complexity (lower bounds of standard problems like searching and sorting). Finally the lectures on real aspects of compiler design and operating system principles were delivered, essentially based, respectively, on the books of David Gries (*Compiler Construction for Digital Computers*) and of Per Brinch Hansen (*Operating System Principles*). The lectures concerning the applied part of the course were delivered by graduate students and were devoted to structured programming in Pascal (based on Jensen and Wirth's famous book *PASCAL User Manual and Report*) and to the design of simple algorithms (syntax analysis, process scheduling, memory allocation etc.). Teaching that course for the first time was a real pleasure for me and for about fifteen years (until a much more specialized and richer curriculum in computer engineering was offered to the students) generations of young Roman engineers underwent that kind of education.

The *Symposium on λ-calculus* took place in Rome in March 1975. It was sponsored by EATCS and hosted by the Italian National Research Council in the recently renovated conference hall. Corrado Böhm was the chairman and the organizer, and being one of the pioneers of λ-calculus studies was visibly proud that the event was attracting the most important experts in the field to Rome. Previously only a few European theoreticians (Wolf Gross, Louis Nolin and Corrado himself) were interested in the subject (which many computer scientists still considered a weird and abstract subject), but more recently, λ-calculus had become a popular topic, especially after Strachey, Landin and Dana Scott had adopted it as a basis for formal approaches to the semantics of programs. Beside the US, Italy, France and the UK, new groups of λ-calculus scholars had appeared in various countries, the Dutch school of Barendregt, Bergstra and Klop above all.

The Rome symposium was the first important event at which, beside the study of λ-calculus as a computation system (Böhm, Klop and Jacopini), λ-calculus was considered from all points of view relevant to the theory of computing: models of λ-calculus based on computing systems (Nolin and Egli), formal approaches to the semantics of programs based on λ-calculus (Welch and Aiello), the relation between λ-calculus and uniformly reflexive structures (Barendregt), non-determinism and parallel computations in λ-calculus (Venturini and Ausiello), notions of subrecursiveness in λ-calculus (Batini and Pettorossi) etc. The invited lecture given by Jaco de Bakker (Fig. 6.5) was a superb presentation of the least fixed point theory of the semantics of recursive programs, in which it was shown how λ-calculus could help to carry out a rigorous treatment of the various parameter passing mechanisms.

The conference was also an excellent occasion for me to host Maurice Nivat for a few weeks. We had foreseen the organization of a summer school in 1976, in which an effort could be made to show the intimate relationship between syntax and semantics of programs, overcoming the apparent dualism (remember that Nivat's approach to semantics was indeed essentially syntactic!). All aspects (title, lecturers, location and organization) had still to be decided, but we had made several assumptions. On that occasion Maurice also announced to me that the first issue of the journal was going to appear soon.

Fig. 6.5 Jaco de Bakker (right) with Henk Barendregt

Indeed, in June 1975, after two years of preparation, the first issue of the journal *Theoretical Computer Science* was finally published (see Appendix L), followed in December by the second issue, and in February and April 1976 by Issues N. 3 and 4. The first editorial board, chaired by Maurice Nivat, consisted of 18 of the world's most authoritative and influential computer scientists: Mike Paterson (associate editor), Jaco de Bakker, Corrado Böhm, Erwin Engeler, Robin Milner, David Park, Arto Salomaa and Arnold Schönage from western Europe, Jiří Becvar and Zdzislaw Pawlak from eastern Europe, Michael Rabin from Israel, Ron Book, Michael Harrison, Richard Karp, Albert Meyer and Jeff Ullman from the US, Andrey Ershov from the Soviet Union and Shigeru Igarashi from Japan. Many of the world's top theoretical computer scientists published papers in the early issues of the journal. The content of the first volume (issues N. 1–4) gives an idea of the topics addressed in the journal and of the high profile of authors: algebraic complexity (Paterson, Schönage, Strassen and Schnorr), developmental languages (Ehrenfeucht, Rozenberg and Skyum), complexity of inherent ambiguity (Savitch), combinatorial properties of codes (Restivo), complexity of Boolean functions (Harper, Hsieh and Savage), minimum storage for stable sorting (Preparata), formal languages (Book, Greibach, Karhumäki, van Leeuwen and Wood) and mathematical foundation of information retrieval (Pawlak). Particularly noteworthy are the papers on the information theoretic lower bound of sorting (Fredman, more than 125 citations), on polynomial time reducibilities (Ladner, Lynch and Selman, 460 citations), on unification in typed lambda-calculus (Huet, more than 780 citations), on call-by-name, call-by-value and lambda-calculus (Plotkin, 1400 citations), and on **NP**-completeness of graph problems (Garey, Johnson and Stockmayer, 2000 citations). Many of these papers belong to the history of the theory of computing. The long journey of the journal, that 40 years later is still happily going on, had started.

Chapter 7
Data Structures, Program Structures

At the beginning of July 1975 I left Rome with my wife and my three children, respectively 4 years, 2 years and 8 months old. Destination Ithaca. The house had been lent to us by a colleague of the Computer Science Department of Cornell University and was on the outskirts of Ithaca, uphill. In Upson Hall (the old building of the CS Department[1]) I was sharing the office with Bernd Reusch, a very friendly and humorous colleague from Dortmund who was doing research on switching theory and linear automata, a particular kind of finite state automata.

Since 1972 Gerry Salton, one of the co-founders of the Department of Computer Science, had replaced Juris Hartmanis as Chairman. Gerry was known as one of the fathers of information retrieval[2]. In 1968 he had published the classic text *Automatic Information Organization and Retrieval* and his research work was now concentrated on a tool for document processing called SMART. Although he was not Chairman anymore, Juris Hartmanis still had a leading role in the department. Every day, at noon, part of the faculty joined him at his table at the Statler Faculty Club. At coffee time the discussion could start and follow the wildest directions: a serious scientific discussion or comments on sport, or a firework of jokes and puns. David Gries provides a nice picture in his speech for Juris' retirement[3]: "Juris chose the topic, and Juris led the conversation. Be it about department business, grilling a recruit, pumping a visitor for news about their department, a Cornell issue,

---

[1] On the occasion of the 50th anniversary of the foundation the CS Department moved to the new Gates Hall, donated to Cornell University by Bill Gates.

[2] Gerard (Gerry) Salton (1927–1995) is credited as the inventor of the vector space approach to information retrieval, a technique that is still used by modern search engines. Both documents and queries are represented by vectors in a high-dimensional space (dimension equal to vocabulary size) whose components are the term counts (or some other measure of term occurrences): the relevance of a document with respect to a query is assumed equal to their similarity, which is given by the cosine between the vector corresponding to the document and the vector corresponding to the query.

[3] http://www.cs.cornell.edu/gries/banquets/jhretireparty/gries.html.

© Springer International Publishing AG, part of Springer Nature 2018
G. Ausiello, *The Making of a New Science*,
https://doi.org/10.1007/978-3-319-62680-2_7

politics, world news, or sports ... No committees! No dreary faculty meetings! Everyone had a voice. There was consensus building, which creates trust, instead of voting, which creates losers and winners and distrust. Everyone won. Juris was the benevolent dictator, and we trusted his leadership and vision because he really did work for the good of the department." When some candidate for a position was visiting the department the real difficult moment in which the decision was made about whether she/he would be hired or not, was not when she/he was giving a talk in the classroom, but rather when they were sitting at the faculty coffee. The tradition started by Juris when he was Chair of the Department was still maintained.

At that time Juris was engaged in the study of properties of **NP**-complete sets, in particular of the hypothesis that all **NP**-complete sets could be isomorphic. In his paper *Juris Hartmanis: Fundamental Contributions to Isomorphism Problems* Paul Young says: "At the time it was widely believed that it was fruitful to consider the relation between **P** and **NP** to be roughly analogous to the relation between the recursive [decidable] sets and the recursively enumerable [semi-decidable] sets. If this analogy were to hold, then from Myhill's work [showing that all semi-decidable sets complete under many-one reductions are recursively isomorphic] it would be natural to believe that all **NP**-complete sets should be polynomially isomorphic." This study was carried out by Juris with his PhD student Leonard Berman and led to proving that all **NP**-complete problems known at that time were polynomially isomorphic, and to supporting the conjecture (known as the Berman-Hartmanis conjecture and still unsolved) that indeed all **NP**-complete sets are polynomially isomorphic[4].

Beside Gerry Salton and Juris Hartmanis several outstanding computer scientists were members of the faculty at Cornell in that moment: Bob Constable, David Gries (author of the already mentioned textbook on compilers: *Compiler Construction for Digital Computers*), John Dennis, Don Greenberg (who developed one of the most important graphic facilities funded by NSF) and, last but not least, John Hopcroft (Fig. 7.1). In 1975 the scientific work of John Hopcroft in a wide range of fields of computer science (formal languages, automata theory, computational complexity and algorithm design) was already impressive. In 1964 he had started his academic career at Princeton where he established a longstanding collaboration with Jeff Ullman. In 1967 John moved to Cornell and, while continuing the research work with Ullman with whom in 1969 he wrote his first important textbook *Formal Languages and Their Relation to Automata*, he started to collaborate with Juris Hartmanis gradually switching from automata theory to compu-

---

[4] If the Berman-Hartmanis conjecture is true, then in practice all **NP**-complete languages would be essentially encodings of the same language under polynomial time bijections. In particular it would imply that there could be no 'sparse' **NP**-complete language A (i.e. a language in which the number of yes-instances of length n grows only polynomially as a function of n) since in that case no polynomial time computable bijection could exist between A and the other known **NP**-complete languages.

Fig. 7.1 John Hopcroft

tational complexity and algorithm design. Let us see how he describes the field of algorithm design in the early 1970s in the Turing lecture that he gave when in 1986, jointly with Robert E. Tarjan[5], he received outstanding recognition "for fundamental achievements in the design and analysis of algorithms and data structures": "During the 1960s, research on algorithms had been very unsatisfying. A researcher would publish an algorithm in a journal along with execution times for a small set of sample problems, and then, several years later, a second researcher would give an improved algorithm along with execution times for the same set of sample problems. The new algorithm would invariably be faster, since in the intervening years, both computer performance and programming languages had improved. The fact that the algorithms were run on different computers and programmed in different languages made me unconfortable with the comparison. It was difficult to factor out both the effect of the increased computer performance and the programming skills of the implementators – to discover the effects of the new algorithm as opposed to its implementation. Furthermore, it was possible that the second researcher had inadvertently tuned his or her algorithm to the sample problems. Conceivably, if the two algorithms were run again on another sample set of problems, the original algorithm might outperform the newer one".

In 1970 Hopcroft had taken a sabbatical year at Stanford and during that time a remarkable collaboration with Robert Tarjan (Fig. 7.2) started and produced outstanding results in the design of efficient graph algorithms for planarity testing and for testing isomorphism of planar graphs. The key ingredient of those memorable results, the technique that permitted an unprecedented degree of efficiency, was the use of depth-first search in the visit of a graph.

Typically, while visiting a graph, two approaches are possible: the so called breadth-first search that consists of exploring, one after the other, all edges that come out of a vertex v before starting to explore the edges coming out of the adjacent vertices, or the so-called depth-first search that consists

---

[5] Robert E. Tarjan is an outstanding American computer scientist. Born in 1948, he took his PhD in Stanford in 1972 under the supervision of Robert Floyd and Donald Knuth. His contributions to the design of efficient algorithms are masterpieces and most of them have become school book material.

Fig. 7.2 From left to right: Robert Tarjan, Al Aho and David Johnson in the early 1970s

of exploring one of the edges that come out of v, reaching vertex w, then following one of the edges that come out of w and proceeding like this until you stop because you have reached a vertex that has already been visited. In that case you go back to the last vertex you have visited (call it vertex u) and take the first unexplored edge coming out of vertex u. When all edges coming out of vertex u have been explored go back to the vertex from which you first reached u.

This procedure, called *backtracking* in games and in artificial intelligence, was indeed crucial to designing efficient graph algorithms. Bob Tarjan remembers: "So I was already thinking about planarity testing. When Hopcroft arrived, we started talking about algorithms and efficiency. Hopcroft came up with an algorithm for finding biconnected components which was sort of a loose sketch to begin with, but what it really was was a depth-first search. I thought about it and tried to make the principle of what was going on in his algorithm more rigorous"[6].

Also the collaboration of John Hopcroft with Karp had produced important results, in particular the classic $O(n^{5/2})$ algorithm for maximum matching of bipartite graphs that was developed in 1973[7]. One of the outcomes of this phase of scientific work was, in 1974, the publication of the important book *The Design and Analysis of Computer Algorithms* by John Hopcroft, in collaboration with Al Aho and Jeff Ullman, which has been the fundamental textbook for generations of computer scientists.

During my stay at Cornell John Hopcroft was working on the study of the reachability problem for Petri nets and (equivalently) for vector addition systems, a problem that was proven decidable six years later in the papers

---

[6] D. Shasha, C. Lazere, *Robert E. Tarjan. In Search of Good Structure*, in *Out of Their Minds. The Lives and Discoveries of 15 Great Computer Scientists*, Copernicus, 1995

[7] Given a graph, a matching is a coupling of vertices such that only vertices connected by an edge can be coupled and each vertex can be coupled at most once. In general, the maximum matching problem asks how to couple the largest possible set of vertices in a given graph; in this case, only bipartite graphs were considered, that is graphs whose vertices can be partitioned into two subsets such that edges only connect vertices in different subsets (hence only vertices from different subsets can be paired).

of Ernst Mayr (who built on a previous incomplete proof by Sacerdote and Tenney) and of Rao Kosaraju, and whose complexity is above elementary functions. He proposed that I look at the problem, but after a couple of weeks I thought that it was wiser for me to turn to some easier stuff. So, for the rest of my stay I devoted my time to the study of programs in 'small' subrecursive classes (in particular 'small' classes of programs for string manipulation).

The study of 'small' classes of programs had become popular after a series of contributions aiming at characterizing classes of programs whose properties could be decidable, and possibly computable in polynomial time. In particular some interest had been raised by the work of Dennis Tsichritzis, concerning the decidability of equivalence for the so-called *simple programs*, i.e. programs based on a limited set of instructions (zero, successor and assignment) and allowing only one level of iteration (corresponding to the first level of the Meyer and Ritchie's LOOP subrecursive hierarchy). My work comprised the study of an analogous class of programs whose set of instructions corresponded to basic operations on binary strings: empty word, two successors (add 1 and add 0 at the end of a string) and assignment.

John Hopcroft also had a rather remarkable PhD student, a young Israeli guy who looked very promising: his name was Zvi Galil. Once he gave a talk on resolution and other techniques for deciding the satisfiability of propositional formulae. I was very impressed and invited him to come to visit us in Rome in 1976. A long-standing friendship started in that moment. As we will see, under various aspects, knowing Zvi had an important role in the research work that we carried out in Rome in the subsequent years.

· Another student of John Hopcroft was Wolfgang Paul. In collaboration with Wolfgang and with Les Valiant, John Hopcroft was working on a very difficult problem: establishing non-trivial bounds in the relation between time and space complexity classes. The problem here was to find out how much space we need to compute all functions that are computable within a given time bound. Of course we need at most an amount of space equal to the number of computation steps, but this is a trivial result. On the contrary, intuition tells us that the amount of space should strictly be less than the number of steps, since space is a reusable resource while time is not. The breakthrough paper *On Time Versus Space and Related Problems* by Hopcroft, Paul and Valiant, in which by means of pebble games on graphs it is shown that space is strictly more powerful than time as a resource for multi-tape Turing machines, was presented at FOCS the following autumn. The paper appeared in the *Journal of the ACM* in 1977 and now counts 224 citations.

Beside Hartmanis and Hopcroft, the other person with whom I was acquainted was Bob Constable who, after the years in which he had been working in abstract computational complexity, had now moved to the field of logic and semantics. At the department, as usual, there were also several visitors. Among others, I remember the visit of Claus Schnorr. Claus was working heavily on the **P** versus **NP** problem, and to the young student who was supposed to pick him up at the bus station, but did not know him personally, Juris came up with one of his jokes. He suggested to the student: "Go to the bus station. When the bus arrives shout '**P** is equal to **NP**' and see who gets back on the bus!". Also Allan Borodin was often visiting the department where he had obtained his PhD and where he had spent one year as a visiting professor in 1975–76. In 1973 Allan had published a book with C.C. Gotlieb *Social Issues in Computing*. This book[8], in which the authors addressed several topics whose impact on our world would become so dramatic (information systems and privacy, computers and planning, computer system security, power shifts in computing, professionalization and responsibility, computers in developing countries, computers in the political process, etc.) had an important role for my work, since social aspects of computing were a hot topic in Italy in the 1970s. Informatics was still regarded suspiciously by some leftist circles because of its apparently negative impact on employment, and the arguments contained in the book could be used as a sort of patent for the 'political correctness' of informatics. The book was also useful when I started to do some consulting for public organizations (e.g. central and local administrations) regarding the design of computer applications with relevant social impacts.

Looking back at that time and at the conversations I had with people in the department, I now realize that in the eyes of Americans, even in university circles, Italy was still seen as a distant country both geographically and socially. News of terror attacks that had taken place in Italy in the early 1970s had given the impression of a destabilized and unsecure country. Probably also rumors of the cholera outbreak in Naples, two years before, had contributed to the depiction of Italy as underdeveloped. Once I was asked whether in Italy one could safely drink water! But overall my stay at Cornell was very pleasant. Juris Hartmanis and his wife Elly, who were familiar with Europe and loved Italy, were very friendly and quite helpful for all practical needs of our family. On the weekends we were often invited by Juris to go to the swimming pool at the club on Cayuga Lake, one of the glacial Finger Lakes that cut through the northern part of New York State, where our little child Sara was cuddled by Juris' daughter Reneta. Juris had a sailing boat there and he loved to spend the weekend on the lake. Sometimes we went out for excursions with Reusch's family and, of course, we once made a trip to Niagara Falls. At the end, on my way back to New York, pleasantly driving on highways surrounded by woods with a gorgeous autumn foliage, I made a stop at the IBM Watson Research Center in Yorktown Heights where I met Arny Rosenberg, who at that time was working on fast access methods for data structures, and I invited him to come to Italy in 1976 to give lectures at the school that I was organizing with Maurice Nivat.

---

[8] Calvin C. (Kelly) Gotlieb was the founder of the Department of Computer Science at Toronto in 1964. On the occasion of the 40th anniversary of the publication of the book a blog was started, and several contributions have marked the role that this textbook had on influencing computer scientists who were more responsive and more ready to understand the impact that computing was going to have on society.

At the end of October I went back to Italy, just in time to start again the lectures of my course on Compilers and Operating Systems, and to resume the research work with the small team of computer scientists at CSCCA. At the end of 1975 the landscape of the theory of computing had indeed changed a lot. The structured approach, which had now become the standard approach to programming and that was producing a variety of programming methodologies mostly based on the concept of top down refinement, had also inspired new ways to address the structuring of data. In particular the algebraic approach to data structures specification was becoming very popular; according to this approach, data items in a program are seen as instances of data types defined, algebraically, in terms of the operations that can be applied to them, and of the properties and effect of such operations. By explicitly specifying the operations that can be performed on data, a stricter control (also at compilation time) can be performed on how data are accessed and modified in a program, to preserve correctness. By defining such operations in an abstract way, their implementation can be maintained separately from their use, providing modularity and easier software maintenance. This is the basis of the currently widely adopted object-oriented approach to programming. The paper by Barbara Liskov[9] and Stephen Zilles *Programming with Abstract Data Types*, which appeared in 1974 in the Proceedings of the ACM SIGPLAN *Symposium on Very High Level Languages* (940 citations), followed in 1975 by the journal version *Specification Techniques for Data Abstraction* (541 citations), had established the basic ground for this approach that would have a very strong impact on future developments of object-oriented programming languages.

Liskov and Zilles' work had its origin on one side in the very early introduction of object orientation in programming languages such as in SIMULA 67, and on the other side in the work that had been started at IBM in the early 1970s by Joe Goguen[10] (Fig. 7.3), Jim Thatcher, Eric Wagner and Jesse Wright (known as the 'ADJ group') with the aim to provide a clean definition of computer science concepts in abstract algebraic terms. In the paper 'Memories of ADJ' that he wrote for the Bulletin of EATCS in 1989 Joe

---

[9] Barbara Huberman Liskov has done outstanding research work that, beside data specification and programming methodologies, included the development of CLU, one of the first object-oriented programming languages, and subsequently covered aspects of operating systems and distributed computing. "For her contributions to practical and theoretical foundations of programming language and system design, especially related to data abstraction, fault tolerance, and distributed computing" Barbara Liskov obtained the Turing Award in 2008.

[10] Joseph Goguen (1941–2006) was an important figure of theoretical computer science. He was a professor at UC Berkeley, UCLA and Oxford University and held research positions at IBM and SRI. His name is essentially related to the early work on the algebraic characterization of abstract data types, but his work spanned a variety of subjects like a categorical approach to fuzzy sets (the topic of his PhD thesis under the supervision of Lotfi Zadeh), the development of a family of programming languages (OBJ) and the introduction of logical concepts in programming.

Fig. 7.3 Joseph Goguen

Goguen says: "In fact I proposed ... the rather grandiose project of providing a uniform, rigorous, abstract treatment of theoretical computer science, similar to what Bourbaki had done for mathematics, but using category theory explicitly, instead of secretly and implicitly, as Bourbaki had done." The two research threads had joined in 1975 when the ADJ group published the paper *Abstract Data Types as Initial Algebras and the Correctness of Data Representation*. In another paper[11] Joe Goguen makes the following consideration on abstract data types: "One of the most important features of modern programming is abstract data types (hereafter, ADT), which encapsulate some data within a module, providing access to it only through operations that are associated with the module. This idea seems to have been first suggested by David Parnas as a way to make large programs more manageable, because changes will be confined to the inside of the module, instead of being scattered throughout the code. For example, if dates had been encapsulated as an ADT, the so called 'year two thousand problem' would not exist. Not all increases in abstraction make programming easier; an abstraction must match the way programmers think, or it won't help. This may explain why ADTs have been more successful than higher order functions".

The concept of abstract data types was immediately adopted by computer scientists. In a few years the idea that a data structure could be seen as an algebra where only pre-specified operations could apply also had a strong impact on pragmatic aspects of programming languages and programming methodologies, and on a variety of applications in which the abstract specification of data could be helpful (for example, a few years later, Jean-François Mascari, Martin Wirsing and I applied the approach to obtain a neat specification of polynomials in algebraic manipulation systems). The most important impact has been in the development of object-oriented languages. Interface, encapsulation and information hiding are concepts born with abstract data types and inherited by the terminology of object-oriented pro-

---

[11] *Tossing Algebraic Flowers*, in *People & Ideas in Theoretical Computer Science*, C.S. Calude Ed., Springer 1998.

gramming. The books by Hartmut Ehrig and Bernd Mahr *Fundamentals of Algebraic Specification* (1 and 2), published by Springer in 1985 and in 1990 respectively, provide a very interesting and complete view of the impact that the algebraic approach had on the development of new programming language concepts and also of the interplay between the algebraic approach to program specification and the other approaches (Scott and Strachey, Hoare, etc.) to program semantics.

The new scenario for theoretical computer science that was appearing in front of us gave Maurice Nivat and myself the inspiration for the organization of the summer course that we had started to conceive: *Data and Program Structures: Syntax and Semantics*. The basic idea was to put together experts in the design of programs and of data structures, and to show the connections between syntactic and semantic aspects. Besides the idea was also to have experts in complexity to analyze how by restricting the expressive power of languages it was possible to cut down computational complexity of the related problems (as happened in the case of programs belonging to 'small' subrecursive classes). The course was meant for young computer scientists, but also for older computer scientists and professionals with limited knowledge of theoretical results who were eager to improve their understanding of the basic concepts of computing.

We also chose the location: the Ettore Majorana Centre for Scientific Culture in Erice, Sicily. The centre had been created in 1963 by the Italian physicist Antonino Zichichi together with an international group of renowned colleagues and had rapidly become one of the major world centres hosting scientific courses and workshops. At the time of the 'Cold War' the Majorana Centre had also become famous for hosting meetings of various organizations in favor of peace and nuclear disarmament. The centre was, and still is, structured in International Schools (now 126!) each one in charge of organizing courses and workshops in one of a variety of scientific domains. In particular the centre also hosted an *International School of Theory and Application of Computers* that was directed by Tom Massam, a physicist living in Geneva. This appeared to us the right framework for submitting our proposal, so I went to Geneva, met Massam and planned with him that our course on 'Programs and Data Structures' would take place in Erice, in May 1976[12].

While we were working on the preparation of the Erice school Maurice invited me to another event that he had organized in Paris, in December, in collaboration with Gérard Viennot: the *Journées algorithmiques*. The interesting aspect of this event was that it represented a joint effort of the two communities of theoretical computer science and combinatorics, and it corresponded to the desire to open new research directions in the world of French theory of computing, mostly dominated, until then, by formal languages and program semantics. These aims clearly appear in the preface of the proceed-

---

[12] An interesting aspect of my trip to Geneva was that for the first (and last!) time in my life I had a dreadful visit to a nuclear shelter that was a few levels below ground in the basement of Massam's apartment.

ings of the meeting that were published by the journal of the French Mathematical Society *Astérisque* (not to be confused with 'Asterix' as was claimed in a recurring joke!): "Ces Journées avaient pour but dans notre esprit de rassembler des chercheurs confrontés à la notion d'algorithme et ce dans les domaines les plus variés des mathématiques et de l'informatique. Nous avons aussi cherché à présenter les résultats les plus récentes de la théorie que les auteurs anglosaxons baptisent 'Complexity of algorithms', subtil mélange de logique (théorie des fonctions récursives), d'algèbre (théorie des langages comme partie du monoïde libre) et de combinatoire. Il s'agit là d'un domaine peu connu en France, puisse ce volume contribuer à son développement".

This declaration reflects how the French community in Paris viewed algorithmic research. In fact, at the *Journées algorithmiques* a variety of aspects of algorithmic research were presented. Beside the classical topics of complexity theory that were addressed by Flajolet and Steyaert (complexity classes and reducibility among combinatorial problems), Paterson (complexity of Boolean functions), Valiant (space-time tradeoffs) and myself (resource bounded approximations of computationally hard logical theories), the other papers were devoted to algorithms in various fields of mathematics. Most of the papers (e.g. Cory, Engeler and Fontet) presented algorithms for problems of an algebraic nature. Schönage and Viennot presented algorithms for combinatorial problems. Rivest and Vuillemin (Fig. 7.4) presented their positive proof of the Aanderaa-Rosenberg conjecture (i.e. any nontrivial property of a graph represented by its adjacency matrix $n \times n$ indeed requires time $\Omega(n^2)$), a result that the authors had already presented at STOC and that appeared to me the most interesting presentation at the workshop. Finally some papers were only indirectly related to algorithms and were essentially formal language papers (e.g. Hotz's result regarding properties of context free languages and Perrot's and Schützenberger's results regarding properties of the free monoid).

But the French theory community was very active in that period in several research directions. At the beginning of February 1976 the French colleagues in Lille, André Arnold and Max Dauchet, started a new conference with a theoretical algebraic flavor: the *Colloque sur les Arbres en Algèbre et en Pro-*

Fig. 7.4 Jean Vuillemin

grammation (CAAP). The new conference reflected, at least at the beginning, the specific French interest in problems on trees seen as an extension of classical problems on strings: characterization of tree languages, tree rewriting systems, tree automata, tree transductions, etc. This type of study was indeed nicely married with the new theoretical results based on abstract data types, and with algorithms on trees and tree-like data structures. Although apparently limited in scope the conference was quite successful and, after the first five editions that were held in Lille, since 1981 when it was organized in Genoa, it started to circulate in Europe with the title *Colloquium on Trees in Algebra and Programming*. Since 1998 CAAP was merged with other theory of programming events in the new ETAPS confederation.

At the beginning of May 1976 I left Rome, destination Erice. Erice is a small town, rich in history, where traces of Greek, Carthaginian, Roman and Norman civilizations meld. It is on the top of Mount Erice, close to Trapani, in western Sicily. Often the clouds accumulate on top of the mountain and the fog brings a surreal atmosphere to the narrow streets of the town. Due to the increase in the number of visitors and also thanks to the many courses that are held in the Majorana Centre, the town has now become richer and frequently full of tourists, but at that time only very few inhabitants were living there and the streets, beautifully paved with polished yellow stone, were almost deserted. When the sky is clear the view from Erice is extremely beautiful. It extends from the island of Mozia to the salt pans around Trapani and, to the south, even to the Tunisian Cape Bon. I was overwhelmed by such an incredible environment. The headquarters of the Majorana Centre are in an old mansion with a large courtyard. To warm the atmosphere between the participants and the lecturers a small barrel of Marsala wine was available in a cellar[13].

At the opening I explained the aim of the course. As was said before, the focus of the course was the relationship that can be established between syntactic, semantic and complexity properties of programs and data structures. The ways in which such relationships could be defined and analyzed were numerous. The first approach that was considered in our course was the subject of the lectures given by Corrado Böhm and this was also the oldest from a historical point of view: attribute semantics, introduced by Donald Knuth in his paper *Semantics of Context Free Languages* in 1968. In this approach to the productions of a context free grammar that control the syntactic structure of a program are associated semantic rules that provide the 'meaning' of

---

[13] In another cellar stood a seismograph. On May 6th it registered the violent shakes of the Friuli earthquake that reached a magnitude of 6.4 and caused more than 900 casualties, particularly in the villages north of Udine. Among the participants in the course was our colleague from Trieste Paolo Sipala who was promptly reassured by his family that no one in the family had been injured. That night Corrado Böhm and I decided that we would propose Udine as the site for ICALP 1978. Paolo Sipala, although he was a professor in Trieste, would be the local organizer since a degree in computer science was not yet open in Udine and no computer science professors were yet there.

the program (in the most simple case, the object code corresponding to the program).

Another, more recent, approach, was based on the theory of program schemes, introduced by David Luckham, David Park and Mike Paterson in 1970 with the paper *On Formalised Computer Programs*. A program scheme is a syntactic abstraction (based on the control structure of a set of programs) that can be transformed into a program by means of suitable interpretations of variables and functions appearing in it. In his lectures Maurice Nivat summarized the results contained in this paper, in particular the fact that two program schemes are equivalent on all interpretations if they are equivalent on the free interpretation (i.e. when they are interpreted in the free magma, corresponding, as we already noticed, to the Herbrand Universe), and extended the result to polyadic recursive program schemes. Also the lectures of Klaus Weihrauch addressed the notion of program schemes in order to introduce the concepts of speed, complexity and optimization of the computation performed by program schemes (in this case monadic program schemes). The lectures given by Gérard Huet were also devoted to program schemes, in which program scheme transformations were studied with respect to classes of interpretations (based on the idea that free interpretations introduced too strong a constraint).

A third way to establish a connection between syntax and semantics of programs was addressed by Rod Burstall[14] (Fig. 7.5) and was based on program transformations. On one side program transformations are related to correctness proofs (the correctness of a program can be reduced to the correctness of a transformed program), on the other side they are related to program synthesis[15]. In the introduction to an extended abstract of his paper Rod Burstall says: "In the lectures I tried to give some insight into recent developments in program correctness proofs, in program transformation and in attempts to synthesise programs from specifications. To what extent do we now understand the reasoning involved in developing computer programs? Is

---

[14] Rodney Martineau (Rod) Burstall is one of the pioneers of British theoretical computer science. He started his career working in operations research and artificial intelligence in Edinburgh in the group of Donald Michie, but his meeting with Peter Landin and Christopher Strachey determined his interest in a mathematical approach to software development. His most important contributions in fact are in semantics of programs, correctness proofs and program transformations. "If one were to summarise the main theme of this work, it lies in the dream that we ought to be able to say in a mathematical language what a program is supposed to do and then prove that it does indeed do this" (*A Collection of Papers and Memoirs Celebrating the Contribution of Rod Burstall to Advances in Computer Science*, Formal Aspects of Computing, 13, 2002).

[15] As is well known, software development is a highly error-prone activity: formalizing program behavior in terms of its effects is thus a fundamental step in order to either prove the correctness of a program (in terms of the correspondence between its behavior and the requirements given) or even to automatically produce this program from a requirements description. Clearly, a suitable formalization of such requirements has to be defined.

Fig. 7.5 Rod Burstall (left) with Robin Milner in the 1980s

our understanding sufficiently sharp to enable us to mechanise this reasoning at least in part?"[16].

The lectures of Ugo Montanari were also devoted to program transformations. Actually Montanari presented an excellent work: respectful of the aim of the seminar he showed how program transformations can be carried out in parallel with data transformations. His lectures[17] were inspired by Liskov's work on the language CLU. Concerning CLU Montanari says: "The idea is to define control and data extensions at the same time, thus achieving a structuring mechanism more adherent to the virtual machine concept especially as far as protection is concerned". Finally program transformations and program synthesis were also addressed in the lectures given by Gianni Degli Antoni, who discussed three different approaches to program synthesis: the informal, intuitive approach, the formal approach, where one starts from a nonprocedural definition of the problem (e.g. in terms of first order predicate logic) and generates the corresponding program, and the inductive approach, where one starts from a sequence of ⟨*input, output*⟩ pairs and generates a program whose behavior is compatible with this sequence.

The lectures of Mariangiola Dezani and of Arny Rosenberg (Fig. 7.6) were devoted to properties of data structures. Mariangiola presented her joint work with Corrado Böhm on data structures associated with generating functions in such a way that operations on the data structures themselves can be replaced by more efficient operations on the corresponding generating functions. Arny Rosenberg's lectures were instead devoted to the work that he had been conducting for a few years at IBM: the study of addressing schemes that could allow direct access to the single elements of a complex data structure (just as we access directly the entries of a matrix or of a table). The subject was clearly explained in the introduction: "Direct access to a data structure requires a coordinate system for the structure. What do the coordinate systems look like? How can they be used when devising storage mappings for a data structure or a family of structures? Where do coordinates come from?"

---

[16] R. Burstall, *Recursive Programs: Proof, Transformation and Synthesis*, Rivista di Informatica, Vol. VIII, Suppl. to n. 1, 1977.

[17] The title of Montanari's lectures was *Data Structures, Program Structures and Graph Grammars*; graph grammars have been Montanari's forte since he published the paper

Fig. 7.6 Arny Rosenberg

The last aspect that was taken into consideration by Hartmanis and myself was complexity. In particular Hartmanis' talk was devoted to illustrating a bunch of results (by Meyer and Stockmeyer, by Fisher and Rabin and by Hartmanis himself) concerning the complexity of decision problems for formal systems (logical theories, formal languages, automata and syntactically defined program classes). The basic fact, common to all such cases, is that even 'small' formal systems may have a very strong expressive power, so that given a Turing machine $\mathcal{M}$ and a string x, there is a formula $\mathcal{F}_{\mathcal{M}}$ in the system that is relatively short (i.e. whose length is polynomial in the length of x) and that has the property to be a theorem if, and only if, $\mathcal{M}$ accepts x in a given amount of time and/or space. This was essentially the argument at the basis of all results concerning the hardness of logical theories and in particular the results presented by Rabin in the already cited IFIP paper. Moreover Hartmanis had shown that any decidable theory 'hides' in itself an infinite number of 'easy' theorems, that is theorems that would be easy to recognize with a specific algorithm (say a finite automaton) but that are extremely hard to recognize with an algorithm that decides the whole theory.

The course went on very smoothly for two weeks. Beside the gorgeous fish based dinners in Erice's restaurants and the evening discussions in the cellar in front of the Marsala barrel, the leisurely part of the program was the excursion to two of the most remarkable Greek archeological sites in Sicily: the temple of Selinunte and Segesta's theater. Although Selinunte's temple is more impressive for the imposing dimensions, Segesta is fascinating because of its location: on top of a hill not too far from the north coast of Sicily, with a fantastic view of the Tyrrhenian Sea. The acoustics of the theater were

---

Separable Graphs, Planar Graphs and Web Grammars on *Information and Control* in 1970!

amazing. All of a sudden, as a tribute to Sicilian culture, Arny Rosenberg standing in the orchestra of the theater, started to play a mouth harp he had just bought in a nearby souvenir shop; all of us sitting on the bleachers could hear the sound perfectly. Another pleasant moment during the course was when I took Juris Hartmanis to see another superb archeological site: the Phoenician island of Mozia. At that time there was no regular transportation to the island but the guidebook suggested what to do. When we arrived in front of the island we started to wave a towel. Juris was rather skeptical but after a while, almost magically, a small rowboat moved from the island 400 meters away and came to pick us up. Juris could not believe his eyes. Another interesting thing that happened during the course is that two Hungarian participants suggested the possibility of having a Hungarian translation of my book on recursive and subrecursive functions. They would take care of finding a translator and a publisher. The proposal indeed was realized. The translation was done by a Hungarian mathematician, Sándor Horváth, and in 1981 my book appeared in Hungarian with the title *Algoritmusok és rekurziv függvények bonyolultsagelmélete*.

We left Erice on May 15 but for many of us the next rendezvous was already established: we would meet soon in two months in Edinburgh at ICALP. The third ICALP congress was organized by Sidney Michaelson[18] (Fig. 7.7) and Robin Milner. The Program Committee was chaired by Michaelson and still consisted of a small group of outstanding computer scientists: de Bakker, Böhm, Book, Burstall, Hoare, Hopcroft, Milner, Nivat, Paterson, Pawlak, Paz, Salomaa and Schnorr. The Preface of the Proceedings, published by the

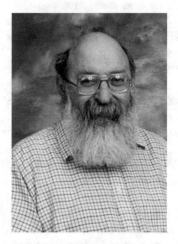

Fig. 7.7 Sidney Michaelson

---

[18] Sidney Michaelson (1925–1991) is remembered as the British computer scientist who first applied computers to the recognition of authors of manuscripts. He founded the Department of Computer Science in Edinburgh in 1966 and was among the first British computer scientists to work on multiuser operating systems.

Edinburgh University Press, is somewhat interesting since it is not just a list of acknowledgements as usual. Rather it reflects the ambition of a scientist, like Michaelson was, who is not satisfied by theory unless he sees answers to technological problems, and contains a scientific appeal to the community: "This Third Colloquium on Automata, Languages and Programming takes place at a time when the technology for constructing information-processing systems has raced far ahead of our ability to theorise about the devices that we can, and do, construct. The need for theories adequate to guide the design and use of such systems grows more pressing every day. It is some comfort that the papers offered to the Program Committee showed that the amount of good research in the Theory of Computation has increased since the last Colloquium but every advance in technology raises more new questions than the theoretical advances have yet answered, and we look forward to future Program Committees being overwhelmed by a flood of worthwhile papers."

Almost at the same time as ICALP an artificial intelligence conference was held in Edinburgh, the *Summer Conference on Artificial Intelligence and Simulation of Behaviour*. My friends Gigina and Mario Aiello were there and we pleasantly spent some hours together walking in the campus speaking about the Italian situation of computer science and about politics (the Red Brigades were becoming more and more threatening). I still remember Mario coughing persistently, the first sign of an evil illness that took him away after a few months.

During ICALP we held 'an informal meeting' of EATCS. About fifty people attended. The meeting was chaired by Mike Paterson (Vice-President) since the President, Maurice Nivat, was absent due to a very bad car accident in which he and Paule had been involved the week before. The future of the association was still unclear and Mike made an effort to encourage the attendees of ICALP to pay the subscription fee (the annual due was 2 pounds!). It was also agreed that in the future EATCS members would be granted a reduced registration fee for ICALP conferences, "the amount of the reduction being comparable to the membership subscription". The potential to offer EATCS members a reduced subscription to the new journal *Theoretical Computer* Science was also discussed, but it appeared that only very few people would have subscribed even at a reduced cost, trusting that their respective institutional libraries would pay the subscription to the journal in any case. Finally, since Maurice wanted to be relieved from the task of editing the bulletin of the association, I accepted to do it, at least initially, for one year.

The bulletin (EATCS Bulletin N. 2) was ready in December 1976 (see Appendix M). Marina Moscarini helped with the editing and we chose as the cover picture an image from one of Paterson's papers on computing the median. We also decided to use the red cover that became the classical cover of the bulletin. In a letter that he sent to me in August, Maurice Nivat enclosed the 'Chairman's letter' that had to be printed in the bulletin. EATCS was still a young, just four years old, organization and the President of EATCS felt the need to renew the motivation for creating the association: "Roughly, the aim of the creators was to provide scientists in this area with a structure comparable to that of SIGACT in the USA: organization of meetings, colloquia, conferences ... and publication of a bulletin containing specific news, announcements and also scientific notes or open problems. We felt such a thing was even more necessary in Europe than in United States for people know little of each other even when working 400 km apart: which researchers from Paris can tell anything of what happens in Leyden, or Colchester or Darmstadt. And how often do we go to fetch the answer to some question as far as California when this answer could be given by a colleague in a nearby university?" He also wanted to stimulate EATCS members to support the bulletin and to provide material to make it richer and more interesting: "For the future of the Association we believe this bulletin is essential: if we can go on, the EATCS will come to adult life. If not we shall remain an informal gathering of friends, almost unknown and always fragile."

The issue also contained the minutes of the 'informal meeting' of EATCS held in Edinburgh, two short research notes by Luc Boasson (*Sur une transduction rationnelle*) and Pavel Pudlák (*Complexity of Mechanized Hypothesis Formation*), reports from research centers (IRIA, Twente and Warwick) and reports on conferences. In particular, de Roever's report concerning MFCS 1976 that had taken place in Gdansk, Poland, beside listing invited lectures and some of the most important talks held at the conference, contained a final comment concerning the relevance of theoretical computer science and illustrating how scientific life in Europe was still hard due to the division into two blocks: "These [MFCS] symposia constitute to my knowledge the only regular meeting place between theoretical computer scientists of eastern Europe and the western world, they are therefore of vital importance for our eastern European colleagues, and constitute for us the only window to the vast eastern European computer science community. No wonder therefore that the organizers of this 5th Symposium seem to be more keenly aware of the present predicament of theoretical computer science than their western colleagues: theoretical computer science, originally motivated by practical problems, seems by and large now to be cut off from programming practice. In itself, this constitutes no objection: this is the normal way in which mathematical specializations develop."

Finally in the bulletin it was possible to find the list of the 15 members of the council and also a complete list of all other 85 members of the association with their addresses.

Chapter 8
Optimization and Approximation

In Rome, during 1976, we had several visitors. Among them, Zvi Galil[1] (Fig. 8.1) had an important role in the development of our research work. At that time Zvi had just recently received his PhD from Cornell University but his visit to Rome had various consequences. First of all, it started a deep relationship that later, when Zvi became a full professor at Columbia, allowed some of the graduates from our university to visit his department and carry on their study there. Second, he pointed out to us the growing interest in a research field that was emerging: the complexity of approximate solution of combinatorial optimization problems.

The issue of finding good approximate solutions for optimization problems, for which an exact optimum solution cannot be achieved by an efficient (polynomial time) algorithm, had been addressed for a long time. For example the performance of heuristics for the well known traveling salesman

Fig. 8.1 Zvi Galil at Columbia on the occasion of the 10th Theory Day (1987). From left to right: Kunsoo Park, Moti Yung, Zvi Galil, David Eppstein, and Stuart Haber

---

[1] Zvi Galil is an outstanding Israeli computer scientist. Born in 1947, he was chairman of the computer science departments first in Tel Aviv and from 1989 to 1995 at Columbia University. Since 2010 he has been Dean of the Georgia Tech College of Computing. His research spans a variety of subjects: string algorithms, graph algorithms, complexity, cryptography, etc. In these fields he has educated various generations of excellent researchers. Following the tradition of Alan Turing, Galil's favorite sports activity is running the marathon.

© Springer International Publishing AG, part of Springer Nature 2018
G. Ausiello, *The Making of a New Science*,
https://doi.org/10.1007/978-3-319-62680-2_8

problem, and for a variety of memory allocation and scheduling problems, had been studied in the 1960s and in the early 1970s by several authors (Lin and Kernighan, Graham, Garey, etc.). But Galil remarked that more recently a few important papers on this subject had stimulated new interest: David Johnson's paper *Approximation Algorithms for Combinatorial Problems* presented at STOC in 1973 (whose journal version was published in 1974 in JCSS and now counts more than 2300 citations), Sahni and Gonzales's paper **P**-*Complete Problems and Approximate Solutions* presented at the IEEE SWAT conference in 1974, and Ibarra and Kim's paper *Fast Approximation Algorithms for the Knapsack and the Sum of Subset Problems* published in JACM in 1975. The situation had changed with respect to previous works: now the notions of complexity introduced by Cook and Karp put the issue in a different perspective, in a more precise context. And the field also started to appear very intriguing. Let us see why.

In his seminal work, Johnson[2] (Fig. 8.2) took into consideration various combinatorial optimization problems for which no algorithm able to find an optimum solution in polynomial time was known. Among the problems considered are, for example, the subset-sum problem (given a finite set of positive numbers and another positive number, called the goal, find a subset whose sum is closest to, without exceeding, the goal), the node cover problem (given a graph find a smallest set of vertices that cover all edges of the graph), and the max clique problem (given a graph find a maximum subgraph with mutually adjacent vertices). Indeed for all such problems the decision version

Fig. 8.2 David Johnson

---

[2] David Stifler Johnson (1945–2016) was an outstanding figure in theoretical computer science. He earned his PhD from MIT in 1973 under the supervision of Michael Fischer. Afterwards he was recruited by Ron Graham at Bell Labs and later became the head of the Algorithms and Optimization Department of AT&T Research Labs from 1988 to 2013. David Johnson is very well known for his large scientific output in the field of approximation algorithms for important problems (such as bin packing and the traveling salesman problem), as well as for laying down the theoretical foundations of approximation algorithms. His book with Mike Garey *Computers and Intractability. A Guide to the Theory of* **NP**-*completeness*, published in 1979, is still one of the major references in the field.

was in the list of problems studied by Karp in his fundamental paper, which appeared in 1972, and classified as **NP**-complete problems. For example the decision version of the *max clique problem* (called the *clique problem* by Karp) consists in deciding whether, given a graph G and an integer k, G has a subgraph G' with at least k vertices that are mutually adjacent. Clearly if any of these optimization problems could be solved in polynomial time, the corresponding decision problem could also be solved in polynomial time, and the outstanding question of whether the two classes **P** and **NP** are equal or not would have a positive solution. Starting from this observation, and therefore from the conjecture that no efficient algorithm for the optimum solution of such problems can exist, Johnson tried to find out whether there exist polynomial time approximation algorithms for any of the problems considered; i.e. algorithms that, although not able to find an optimum solution, in polynomial time can at least find a solution which is (guaranteed to be) not too far from the optimum.

But actually Johnson's paper, although rich in examples of approximation algorithms, had raised more questions than it was able to answer. In fact, the problems considered by Johnson, although their decision versions were all **NP**-complete and therefore in some sense equivalent from a complexity point of view, exhibited a deeply different behaviour from the point of view of approximability. For example, for the subset-sum problem there are algorithms that in any case can achieve a solution which is arbitrarily close to the optimum, while for the node cover problem it was easy to find an algorithm able to guarantee a solution of cardinality at most twice the optimum, but apparently no better approximation could be obtained. Finally for other problems it appeared that, for any k, no polynomial time algorithm could guarantee a solution at most k times worse than the optimum. This had been proved, for example, for the traveling salesman problem by Sahni and Gonzales in their 1974 paper: any polynomial time approximation algorithm for the traveling salesman problem would allow the solution of the Hamiltonian circuit problem in polynomial time (a very unlikely event since the Hamiltonian circuit problem, which is the problem to decide whether, given a graph, there exists a tour that goes exactly once through all vertices of the graph, is **NP**-complete).

In any case the field of approximation definitely appeared worthy of being explored in all its extent. First of all, since complexity theory had been initiated as a theory regarding languages and decision problems, it was still not clear how to extend complexity concepts to optimization problems. Although Johnson called these problems 'polynomial complete optimization problems' a formal definition of their complexity was yet to be provided. Also, a precise characterization of the classes of problems that exhibited different approximability properties was still missing. In some cases the proof of bounds to the quality of achievable approximation, or the proof that no constant approximation at all could be achieved in polynomial time (unless all **NP** problems could be solved in polynomial time), could be easily obtained. For example it was easy to prove that (unless $\mathbf{P} = \mathbf{NP}$) graph coloring could not be better approximated than with a ratio 4/3. But for other problems (e.g. the max clique problem) there was a total lack of arguments that could lead to the discovery of polynomial time approximation algorithms, or to prove that they did not exist. Moreover it was necessary to extend the notions of reducibility and of completeness to optimization problems, and possibly to identify those reducibilities that could preserve good approximability properties. Johnson himself had raised the question: "What is it that makes algorithms for different problems behave in the same way? Is there some stronger kind of reducibility than the simple polynomial reducibility that will explain these results, or are they due to some structural similarity between the problems as we define them?" All these issues and open questions stimulated various research groups. Among others, Azaria Paz in Israel and our group in Rome.

At that time I was collaborating with Marco Protasi and two other young post-docs: Alessandro D'Atri[3], who had done his master's thesis with Paolo Ercoli developing a categorical approach to the representation of encodings between information structures, and Alberto Marchetti-Spaccamela[4] who had done his master's thesis in operations research. Both had decided to move to graph algorithms and optimization and had joined my group. We started to work on some of the issues I mentioned before and at the next ICALP in Finland, and MFCS in Czechoslovakia, we were able to present our ideas. Actually it now appears that the way to obtain a satisfactory answer to the questions concerning reductions between optimization problems and completeness in approximation classes was still very long. In fact, a more precise and comprehensive picture of the theoretical aspects of approximation algorithms was achieved only at the end of the 1980s and the beginning of the 1990s, in particular thanks to the work of Orponen and Mannila (*On Approximation Preserving Reductions: Complete Problems and Robust Measures*), Papadimitriou and Yannakakis (*Optimization, Approximation, and Complexity Classes*), and Crescenzi and Panconesi (*Completeness in Approximation Classes*). Finally were the breakthrough papers by Feige, Goldwasser, Lovász, Safra and Szegedy (*Approximating Clique is Almost NP-complete*, almost 500 citations) and by Arora, Lund, Motwani, Sudan and Szegedy (*Proof Verification and Intractability of Approximation Problems*, more than 2200 citations) that in 1991 and 1992 respectively explained why for some optimization problems (e.g. max independent set, max clique and min graph coloring) finding an approximate solution may be as hard as finding the op-

---

[3] With Alessandro D'Atri I shared a lot of fruitful research work over several years, until 1987 when he took a full professor position in L'Aquila and we lost the opportunity to work together. In 1997 he became a full professor at LUISS (the Free International University of Social Science "Guido Carli") in Rome, where he created the Research Center of Information Systems. He died in 2011 when he was only 60 years old.

[4] Alberto Marchetti-Spaccamela, born in 1954, was first a professor at L'Aquila University and from 1990 he is a professor at Sapienza University of Rome. For his excellent scientific production in the field of analysis and design of algorithms he is among the top European computer scientists.

timum one[5]. Still our results and the work that other groups carried out in those early years can be recorded among the first attempts to extend the notion of reducibility between decision problems to a new notion of reducibility between optimization problems.

The winter of 1977 had been rather troubled at our university. Antonio Ruberti, the former director of my institute, had been elected rector of the university in 1976 and it was his first year in charge. It was not easy to be the rector of one of the largest universities in the world (about 120,000 students) in a period of social unrest. Various groups of students of different political orientation, ranging from the extreme right to the extreme left, had almost daily physical confrontations or even clashes. In February Luciano Lama, the leader of the major union (CGIL, the workers confederation that was politically related to the Communist Party) was supposed to speak at the University of Rome. However, the extreme left students, who were fighting against the support that the Communist Party was giving to the Christian Democrat government and against the laws that the Ministry of Education was proposing to the parliament, did not allow him to speak and literally chased him out of the campus. This was one of the worst events that happened during that year, but on top of this, always hanging, was the presence of the Red Brigades that in various assaults injured or killed industrial managers, unions leaders and, sometimes, university professors.

Teaching a course at the university in that atmosphere was not easy either. Often, students decided to suspend the lectures to organize a rally, and sometimes the buildings were occupied, but altogether the situation in the Faculty of Engineering was not so bad. The students were generally highly motivated and that year I had a few exceptionally good students (at least four of them ended up as professors in various universities). In any case, thanks to his exceptional scientific leadership, Ruberti's mandate as rector, which lasted until 1987 when he became Minister of Scientific and Technological Research, is still remembered as one of the best periods for our university, that just in those years recovered its medieval name: 'La Sapienza'.

In 1977 ICALP was held in Turku, Finland. The year before, in Edinburgh, it had been decided that ICALP should become an annual event and our Finnish colleagues had proposed to hold the conference in Turku, where there was a major Finnish theoretical computer science research group, led by Arto Salomaa[6]. The conference was organized by Arto and he, together with

---

[5] Actually both papers were based on a characterization of **NP** in terms of probabilistically checkable proofs (known as **PCP** Theorem). For the history and development of the **PCP** Theorem and its consequences see http://courses.cs.washington.edu/courses/cse533/05au/pcp-history.

[6] The name Arto Salomaa has appeared several times already in these notes. Arto is one of the most prominent and prolific mathematicians and computer scientists in Europe. Born in 1934, he has published more than 400 scientific papers and has authored or co-authored 46 textbooks. During his career he has addressed several research topics, mostly related to the field of automata and formal languages (a domain to which he was introduced in Berkeley by John Myhill in 1957). Among his students are some of the most prominent European

Magnus Steinby, was also the editor of the proceedings. From 1977 until 2015, all ICALP proceedings were published by Springer; the proceedings of the Turku conference were LNCS Vol. 52.

This was the fourth ICALP and the conference was gaining momentum, with 43 contributed papers. Although algorithms and complexity issues were addressed by several speakers, the majority of the papers (24) concerned formal languages and semantics. Five papers were still devoted to Lindenmayer systems. Among the complexity papers, I think one of the most relevant was the paper by Wolfgang Paul and Bob Tarjan providing bounds in space-time trade-offs, based on pebble games on directed acyclic graphs, a follow up of the work that Wolfgang Paul had done the year before with his advisor John Hopcroft and with Les Valiant. But for me, of course, the paper that appeared more interesting was the one presented by Azaria Paz and his student Shlomo Moran: *Non-deterministic Polynomial Optimization Problems and Their Approximation*. The paper addressed the same area as our paper at the conference *On the Structure of Combinatorial Problems and Structure Preserving Reductions*: the complexity of optimization problems and the theoretical aspects of approximation. In particular Paz and Moran's paper analyzed the complexity of those optimization problems whose associated decision problem was **NP**-complete, and provided necessary and sufficient conditions for their approximability. Although the conditions were not easy to test, the result was a big step forward in the study of approximability of hard optimization problems.

My memories of ICALP 1977 are also related to one of the figures who has marked the history of theoretical computer science in Europe: Marcel-Paul Schützenberger. Schützenberger attended ICALP in Turku accompanied by his son Mahar (who died in a car accident only three years later). On that occasion I had, for the first time, the chance to talk to him and to listen to his inspiring views on the role of mathematics in the theory of computing, the only aspect of computer science that he thought worthy of being studied. Also the figure of Arto Salomaa, scientifically and physically imposing, is still impressed in my mind.

During ICALP we had an EATCS Council meeting and the second General Assembly of the Association. At the council Maurice Nivat told us that he had decided to step down as EATCS president, and he proposed that Mike Paterson should be elected as the new president. Of course the election was unanimous. Besides, Maurice Nivat and Jaco de Bakker were elected as vice-presidents. Grzegorz Rozenberg and Hermann Maurer (Fig. 8.3) became new Council members, and agreed to serve as treasurer and secretary respectively.

---

computer scientists: Neil Jones, Juhani Karhumäki, Jarkko Kari, Lila Kari, Mogens Nielsen, Sven Skyum, and Paul Vitanyi. Arto has been one of the most active members of EATCS; he became president of the association in 1979 and remained in charge until 1985.

Fig. 8.3 From left to right: Hermann Maurer, Arto Salomaa, Derick Wood (the MSW group) in the 1970s

Also the first non-European member was elected to the council: Ron Book[7] from UC Santa Barbara (USA).

Hermann Maurer and Grzegorz Rozenberg have both been leading figures since the early years of theoretical computer science, both very prolific and energetic. Both have greatly contributed to the development of the European theory community. Maurer, born in 1941 in Austria, became a professor at Karlsruhe in 1977 and since 1978 he has been a professor at Graz. His scientific output in the fields of formal languages, Lindenmayer systems, automata theory, and data structures has been huge. Particularly remarkable are the 'MSW papers', 30 papers regarding formal languages and L-systems that he wrote together with Arto Salomaa and Derick Wood between 1975 and 1981. Among his students are some of the brightest European computer scientists (e.g. Emo Welzl and Herbert Edelsbrunner)[8].

Grzegorz Rozenberg is also extremely prolific and energetic; his production includes more than 500 scientific publications and several books. He was born in 1942 in Poland and received his PhD in mathematics in Warsaw. Since 1979 he has been a professor in Leyden and an adjoint professor in Boulder, Colorado. His research interests span a wide variety of fields: formal languages, L-systems, automata theory, graph grammars and graph transformations, Petri nets, concurrent systems, and natural computing[9]. Grzegorz became the EATCS Bulletin editor in 1981 and was subsequently EATCS President from 1985 until 1994. In that period (in particular in 1989 when I organized ICALP in Stresa together with Mariangiola Dezani) we were more

---

[7] Ronald Vernon (Ron) Book, was one of the main scientists who laid the foundations of formal languages and automata theory in the US. He died in 1996 when he was only 60 years old.

[8] Having been one of the European leaders in theoretical computer science, Maurer moved to more applied research activity (networks, graphics etc.) and collaboration with industry. He is also known as a science fiction author.

[9] Grzegorz Rozenberg is defined by Wikipedia as the 'guru' of natural computing and in fact he has promoted the field in several ways (in particular founding the International Journal of Natural Computing and the Series C of the journal Theoretical Computer Science: Theory of Natural Computing). Rozenberg is also known as a professional magician, under the pseudonym Bolgani.

frequently in contact, and I have been honored to receive his warm friendship. Grzegorz has been very deeply and wholeheartedly involved in EATCS
activity. He says: "My involvement with EATCS falls under the cliché of the
exemplary military career: from soldier to general. ... Bringing together scientists from various countries and various disciplines of science is for me a
noble cause, certainly worth devoting a considerable part of my life."[10]

At the General Assembly in Turku our proposal to hold ICALP 1978 in
Udine, Italy was also presented and approved. In view of our project, Corrado
and I (the supposed co-chairs of the conference) were careful to consider all
aspects of the organization, wondering whether we might be able to do a
similarly good job. The traditional Wednesday excursion was a picnic place
in the woods, close to the sea. Of course there was also a sauna cabin and
many of us did not want to miss a classical Finnish experience. After the
sauna everybody jumped into the sea which, despite the season, for me and
other Mediterranean guys was pretty cold. While I was trying to get dry
under a pale sun Hermann Maurer approached me. He told me that he could
take care of editing the EATCS Bulletin if I agreed. Of course I was happy
to be relieved of that task and I gladly agreed. By the way, that was a good
decision because Hermann did a wonderful job editing the bulletin for four
years, from issue N. 3 (October 1977, when for the first time the now famous
logo of EATCS appeared on the cover) to issue N. 14 (June 1981). He also
introduced the publication of three issues per year, and brought the bulletin
to the current rich scientific content and professional editorial level. Bulletin
N. 4 included photos documenting ICALP 1977 and MFCS 1977. Bulletin N.
11 was 180 pages thick, and Peter van Emde Boas (Fig. 8.4) had become the
official pictures editor.

The bulletin was improved even more when Grzegorz Rozenberg (Fig. 8.5)
started to edit it in 1981, introducing more sections and more photos[11].

Fig. 8.4 Peter van Emde Boas (left) with Ivan Havel

---

[10] *The Magic of Theory and the Theory of Magic* in *People & Ideas in Theoretical Computer Science*, C.S. Calude, Springer, 1998

[11] Grzegorz also introduced in the bulletin the drawings by his son Daniel, the well known
Dutch artist Dadara, whose pseudonym is reminiscent of the great Italian designer of
graphic novels, Milo Manara. The history of the EATCS Bulletin is described in Ute and
Wilfried Brauer's report *Silver Jubilee of EATCS*, Bulletin of EATCS, N. 62 (June 1997).

Fig. 8.5 Grzegorz Rozenberg (left) with Cris Calude

For me the trip to Finland was concluded by a visit to Leningrad (today Saint Petersburg), with long lazy strolls along the Nevsky Prospect, where visions from Gogol's and Dostoevsky's novels finally became real. Unfortunately that was to be my first and only visit to Russia.

Back in Rome I started preparing the two talks that I was supposed to deliver at MFCS: one invited talk on the structural aspects of **NP**-complete problems and their associated optimization problems, and one submitted contribution. In 1977 (odd year) MFCS was held in Czechoslovakia, in Tatranska Lomnica. This was the third time I had visited Czechoslovakia and I started to be acquainted with that lovely country.

After MFCS 1973, I visited Prague for one week in January 1976, invited by Jiří Becvar. Those seven days in Prague have been very important in my life. Beside giving a few talks at Charles University, I spent most of the time discussing research topics with Michal Chytil, a brilliant young researcher I had met at ICALP. At that time Michal was working in formal languages and computational complexity. In particular he was interested, as I was, in the properties of simple programs on strings. In his company I also had the chance to visit the city in a more intimate way. After this visit, during which I enjoyed long pleasant walks in the narrow streets of Prague, felt the mystery of the Jewish cemetery, saw the statues of Charles Bridge covered in snow, visited several (now disappeared) antique bookshops, had joyful dinners with Michal and other colleagues in Charles University in the many small restaurants hidden in cellars, I fell in love with Prague. I was profoundly overwhelmed by its magic atmosphere, so well described in the book *Praga magica* written in 1973 by Angelo Maria Ripellino, a poet and the greatest Italian expert on Slavic literature. Prague was now competing with Paris as my second most beloved city (the first being, of course, Rome).

So when the time of MFCS arrived I was very glad to return to Czechoslovakia. Marco Protasi and I decided to go by airplane to Vienna, then we took a bus to Bratislava and finally a train from Bratislava to Kosice. We had not realized that the police control at the border on the bus was much more annoying when traveling by road, than if we would have been traveling by air or by rail. In fact, at the border our luggage was carefully searched and we were astonished when a policeman asked us to explain the meaning

of the drawings on our slides (schemes describing relationships among classes
of isomorphic optimization problems).

In any case when we arrived in Tatranska Lomnica we were captivated by
the fantastic setting. The location, in the woods of the Tatra mountains, was
extremely beautiful. The conference had been organized by Józef Gruska[12]
(Fig. 8.6) who was also editor of the proceedings. At that time Gruska was
affiliated with the Computing Research Centre of the United Nations Development Programme in Bratislava, that was also the main sponsor of MFCS.
Józef was one of the pioneers of formal languages research in Czechoslovakia
and he had an impressive record of papers devoted to various aspects of
context free languages. In particular he was interested in the descriptional
complexity of formal languages and its relationship with computational complexity (essentially, as the expressive power of a language becomes stronger,
determining its properties becomes harder: a subject that was strictly related
to the work of Hartmanis mentioned in the previous chapter). The conference
was as usual organized with a series of invited talks (15) in the morning and
contributed talks (46) in the afternoon.

Again the conference revealed its strength in allowing the encounter between East European and West European computer scientists. Here I met,
for the first time, some of the outstanding leaders of the theory of computing in eastern Europe: Józef Gruska (of course) and then Andrzej Salwicki

Fig. 8.6 Józef Gruska

---

[12] Józef Gruska, born in 1933, is one of the founding fathers of Czech theoretical computer
science. His interests range from formal languages to parallel computing and quantum
computing (not to speak about his collection of Nativity sets!). Since the 1990s he has
been a professor at the Masaryk University in Brno. He has promoted several initiatives
including the MFCS and SOFSEM series of conferences. After a long fight against the other
technical committees that wanted to maintain their share of theoretical issues, with the
help of the 'pro-tempore' IFIP President Blagovest Sendov, Gruska was able to convince
the federation to create the Specialist Group of Foundations of Computer Science (that
later gave rise to IFIP TC1).

from Warsaw, working in algorithmic logic with his wife Grazyna Mirkowska (with whom a longstanding friendship was established) and Gerd Wechsung from Jena, working in computational complexity, who invited me to visit his excellent group of researchers in 1978. I also met with Sándor Horváth who was translating my book in Hungarian, and who also had proposals on how to improve the presentation in various parts of the book.

The leisurely part of the conference is also memorable. Hiking in the clean and cool air in the High Tatras was an extreme pleasure. I also remember having excellent dinners in some of the small restaurants hidden in the woods.

Soon after the conference I went back to Rome but many of the MFCS participants were going to Poland where another event was taking place. A new series of conferences in the field of the theory of computing was starting: the international conference on *Fundamentals of Computation Theory* (FCT). The first FCT, in 1977, was organized in Poznań by Marek Karpinski[13] and attracted a great number of excellent papers. This series of conferences still goes on today, with a biennial frequency.

In Italy, in 1978, the political situation was indeed worsening. The winter had been dreadful and gloomy. The Red Brigades attacks intensified and on March 16 with the abduction of the Christian-Democrat leader, Aldo Moro, on the day of the first speech of the newly appointed Prime Minister, Giulio Andreotti, to the Parliament, and the slaughter of his body guards, reached their climax. On May 9 Moro was found dead in the trunk of a red Renault car.

Nevertheless we had to proceed with the organization of ICALP 1978. By the submission deadline (November 15 was traditionally the submission deadline for ICALP, before the EasyChair age) we had received about 90 papers and we started to make copies to send out to the Program Committee members. Besides I made a couple of trips to Udine to choose the venue, to establish contacts with hotels and, the most pleasant part of the survey, to check the quality of the restaurants. Fortunately the local support was extremely efficient. It was provided by the *International Centre for Mechanical Sciences* (CISM), whose main activity consisted of organizing courses, workshops and conferences in various domains of engineering, mainly mechanical engineering, and computer and systems sciences. The seat of CISM was in a wonderful building from the 15th century that had been renovated several times, but still had a wonderful classroom with ancient wooden seats and desks, and a hall with frescos on the ceiling and the coats of arms of ancient Udine families. The people of CISM were very kind and helped us to find a suitable place for hosting ICALP lectures: the 'aula magna' of a technical high school. As I said a university had not yet been created in Udine. In-

---

[13] Marek Karpinski, born in 1948, is a great and very productive Polish mathematician and computer scientist. His research work has been devoted to a large series of subjects in algorithm design, discrete mathematics, optimization, and computational complexity. Karpinski is a professor at Bonn University and has also been a professor at Berkeley and Princeton.

deed it would be created a few years later as one of the measures adopted to reconstruct the city after the earthquake.

In February the Program Committee met in Paris to select the papers. The meeting was held in Nivat's office at Jussieu and the discussion was very animated, as it usually was when physical PC meetings were held. Finally 37 papers were selected. In addition we had invited a couple of speakers.

The proceedings were edited by Corrado and myself. For the first time the proceedings carried the EATCS logo on the spine. An interesting comment in the preface shows how slow the evolution of topics presented at ICALP in that period was: "In addition to the main topic treated – automata theory, formal languages and theory of programming – other areas such as computational complexity and λ-calculus are also represented in the present volume." This statement demonstrates that although new topics were starting to appear, the 'core business' of ICALP was still related to the traditional three fields of automata theory, formal languages and the theory of programming. Actually 18 papers presented in Udine belonged to this 'core', but 11 papers were in the field of algorithms and complexity, four in λ-calculus, two in the area that we would today call distributed computing and one addressed the subject of inductive inferences. The conference program still included six papers in French; for which a special session was arranged.

The invited speakers were David Harel and Frank Harary. Harary (who died in 2005 at the age of 84) was at that time a professor at the University of Michigan. He was one of the fathers of graph theory and one of the founders of the Journal of Graph Theory. He was extremely prolific (he wrote about 700 papers in his life) and was very proud to have given hundreds of lectures around the world. When I went to the train station to pick him up he soon showed me his famous alphabetical list of all places where he had given lectures, happy for the chance to add Udine to it. Unfortunately his lecture is not included in our proceedings.

Among the most relevant contributions at ICALP 1978 was definitely Piotr Berman's paper *Relationship Between Density and Deterministic Complexity of $\mathbf{NP}$-complete Languages*, in which it is formally proved that if there is an $\mathbf{NP}$-complete language over a single letter alphabet then $\mathbf{P}$ would be equal to $\mathbf{NP}$, a result that was strictly related to the (Leonard) Berman-Hartmanis conjecture on the isomorphism among $\mathbf{NP}$-complete problems that was cited in the previous chapter. Another paper that has been rather popular for some time (and now counts more than 300 citations) is the paper by Jean-Jacques Berry *Stable Models for Typed Lambda-calculi*, in which, following Plotkin and Milner's criticism on Scott's models, a stronger type of models was introduced.

The conference went smoothly with a very warm and relaxed atmosphere. This was the first time I had organized an international conference and it was really challenging to guarantee the quality of both the scientific part and of the social aspect. Indeed it was a real pleasure to see so many colleagues and friends gathered in Udine. Beside the Italian colleagues, like Antonio

Restivo, Alberto Apostolico and Mariangiola Dezani, it was nice to meet again Zvi Galil, Jean Eric Pin, Mike Paterson, Maurice Nivat and Azaria Paz. In particular I remember with pleasure the warm feeling that Azaria and his wife Erela established with my little daughter Sara, only three years old. In Udine I also met for the first time Manfred Kudlek, a professor in Hamburg, specialized in formal languages. Kudlek became famous in the EATCS community as the only person who attended all ICALP conferences, until his death in 2012 at the age of 72.

Udine is a nice city with an Austro-Hungarian look. Wine marks the character of the city, actually it is in the character of the whole Friuli region. The typical way in which the people of Friuli start their day is by drinking 'un'ombra di vino', literally a shadow of wine, in practice half a glass of Friuli tocai or traminer, rather than a cappuccino. The most daring start with a small glass of 'grappa'. Several pubs and wine bars in the city center provide warm, convivial hospitality for small parties. One of the places where the participants of the conference more frequently used to meet was called *Spezieria pei sani* (pharmacy for healthy people!). During the excursion we took the participants to the most beautiful archaeological site in the region: Aquileia, an ancient Roman city that was abandoned as a consequence of the destruction perpetrated by Attila. The floor of the 4th century cathedral is one of the largest and best preserved Roman mosaics.

The traditional Thursday evening social dinner was a great success, not only because of the excellent food, but also due to the unexpected show by Mike Paterson. In order to entertain the participants at the banquet I had asked Mike to deliver a speech. Actually I had in mind a short (but serious) scientific talk, but Mike decided to give an amusing phony speech. The title referred to a generalization of generalized sequential machines (themselves a generalization of finite state machines), and the talk made fun of the typical abstract style of applications of category theory in computer science. The algebraic definitions of these super abstract machines required dozens of mappings, and the slide presentation consisted of dozens of overlapping, intertwining and folding plastic foils. It was fantastic, and also somewhat provocative toward the algebraic approaches that were prevailing in some areas of theoretical computer science, from program semantics to abstract data types.

In conclusion we were so happy about the development of the conference, the support provided by CISM and the hospitality we had found in the city that, together with Marco Protasi and Alberto Marchetti, we decided to organize another event in Udine. Again with the support of CISM, we ran a summer school on algorithms for optimization problems in 1979.

After ICALP, Maurice Nivat, who was traveling with his wife Paule and his son Jean-Luc in a Ford Transit, came to visit us in Dogliani. He was enchanted with the old 18th century house, and I still have the drawings of the house that Maurice and Jean-Luc painted, sitting on the lawn in a lazy afternoon.

Then, at the beginning of September, MFCS was held in Zakopane, again in the Tatra mountains, but this time on the Polish side. I decided to go. I had been invited to Jena by Gerd Wechsung and I thought that it would be a good idea to combine the two stops in the same trip.

Zakopane is a famous resort for skiing and hiking, and the city had, so to say, an alpine aspect. The conference had been organized by Józef Winkowski, with seven invited talks and 44 contributed talks. The invited talks were totally oriented toward logic and semantics: Milner and Dijkstra addressed aspects of semantics of concurrency, Mayoh and the ADJ group (Wagner, Thatcher and Wright) presented algebraic approaches to programs and data structures, and Courcelle and Nivat presented their algebraic semantics for recursive program schemes. The contributed papers were also mostly in the field of semantics and of algorithmic logic, the research field that had been developed by Salwicki and his group, with the notable exception of a few papers that addressed two emerging new areas that would explode in the 1980s: the theory of relational databases and the theory of distributed systems. In the first group there was a paper by François Bancilhon[14] (who in the meantime had moved from automata theory to database theory) on the completeness of query languages, a paper that became quite famous, and a tutorial by Jorma Rissanen that also gained some popularity. In the second group there was a paper by Francez, Hoare and Lehmann, with a proposal for extending denotational semantics to nondeterministic and concurrent communicating sequential processes, plus a series of papers by several authors devoted to Petri Nets. Only five papers were devoted to algorithms and complexity.

Since I had started to study some aspects of the design of relational databases, the contact with François Bancilhon and his colleague Nicolas Spyratos, both at IRIA, was very important. Besides during the excursion to the mountain Józef Gruska invited me to visit his group in Bratislava next January, and to give some lectures on the complexity of optimization problems.

During 1978 several other interesting initiatives took place in Europe. During the summer of 1978, just a few days after ICALP, another NATO summer school was held in Marktoberdorf, organized by Fritz Bauer and Manfred Broy. The title was 'Program Construction' and the impressive roster of lecturers included Edsger Dijkstra (program conception), David Gries (program verification) and John Guttag (abstract data types). Indeed the series of lectures given by Dijkstra was remarkable. The title was *The Thinking Programmer* and the lectures were devoted to "the interplay between invention and formal techniques". The emphasis was all on the concept of 'invention' as the following quote explains: " ... in this series of lectures I

---

[14] François Bancilhon is an example of a theoretician turned entrepreneur. After first working on automata theory Bancilhon started to do research in database theory and then, after working for a few years in the US, decided to found a company, O2 Technology, devoted to the production of object-oriented databases. Since then Bancilhon has continued his activity in various hi-tech companies both in France and in the US.

would like to appreciate the gadget (i.e. the computer) as the embodiment of an intellectual challenge that is also without precedent, viz. the challenge to program the gadget. This challenge seems unique in the combination of the possibility for unmastered complexity – programs are among the most complex things ever conceived – and the ultimate, but misleading, simplicity of a world of zeros and ones alone".

Besides, 1978 also marked the birth of another famous workshop: the *International Workshop on Graph Grammars and their Application to Computer Science and Biology* (which after a while was commonly known as *Gra-Gra*). It took place for the first time in Bad Honnef at the end of October 1978, and was organized by Volker Claus, Hartmut Ehrig, and Grzegorz Rozenberg. In that period graph grammars were a ubiquitous concept in computer science. In the preface of the proceedings it states: "... in many applications of formal language theory applying string languages was considered to be the first step only, leading to a more general theory where sets of multidimensional objects, as well as various transformations of them could be (finitely) described and studied. Then for example various problems in databases, semantics of programming languages, two-dimensional programming languages, data flow analysis or incremental compilers call for finite grammatical definitions of sets of graphs, whereas various problems in picture processing and biological pattern generation require finite grammatical (or machine) descriptions of sets of maps". The proceedings start with two excellent surveys, one by Hartmut Ehrig, regarding "the basic constructions and results which have been developed in the algebraic theory of graph grammars up to now" and one by Manfred Nagl containing a very exhaustive bibliography on graph grammars. Most of the authors that had made use of graph grammars in their studies were there, including Montanari, Kreowski, Lindenmayer, Mayoh, and Paz. In particular among the authors were my colleagues from Rome, Carlo Batini and Alessandro D'Atri (who had been encouraged by Paolo Ercoli to do research in the field of databases), who presented their methodology for relational database design based on graph rewriting, an approach that in Italy would have several theoretical and practical developments in the 1980s.

The workshop, whose creation shows how graph rewriting systems were considered relevant in computer science, was competing with another well established workshop partly covering a similar domain: the *Workshop on Graph-Theoretic Concepts in Computer Science* (known as WG), which started in Germany in 1974 as *Fachtagung über Graphentheoretische Konzepte der Informatik*, but in the subsequent years had become a truly international event, although mostly held in Germany[15].

After MFCS I took a train from Krakow to Jena. Jena was, and still is, an industrial city. The city center had been heavily rebuilt after the war and when I visited, in September 1978, the most important landmark was the Jen Tower, the tower that hosted the University of Jena (the Friedrich Schiller

---

[15] On some occasions when the two events coincided, the competition even materialized into a soccer match between teams from the two conferences.

University)[16]. A section of the city was occupied by Soviet troops and their families, and this gave Jena a rather gloomy aspect. On the contrary the countryside was very lively and delightful as I had the chance to see during various excursions. Thuringia is a very beautiful region and is very interesting from the touristic and historical points of view. In a series of excursions with Gerd Wechsung I had the chance to visit the fields where the battle of Jena was fought and also the two Bach cities of Erfurt, the capital of Thuringia where Bach used to come for family meetings and where he played on the organ of the Augustine monastery, and Eisenach, Bach's birthplace.

The group of young researchers that had been gathered by Gerd Wechsung (Fig. 8.7) was extremely active and interesting. Their work in structural computational complexity started to be very fruitful. Among them, Klaus Wagner and Andreas Brandstädt have become very well known computer scientists: Wagner continued his work in computational complexity and only recently moved toward game theory, while Brandstädt moved from computational complexity to graph algorithms in the 1980s, and now has a huge production concerning a variety of graph problems. Gerd's work concerned complexity measures and oracle computations, but he was very interested in all results and approaches that were emerging in the field, including the most recent work in complexity of optimization problems. He was collecting material for a book that he would write together with Klaus Wagner in the early 1980s: *Computational Complexity*, one of the first books that covered all aspects of complexity in a kind of encyclopedia.

Fig. 8.7 Gerd Wechsung (center) with Boris Trakhtenbrot (left) and Jörg Rothe (right)

Mostly due to Wechsung's work, the University of Jena had become an attractive place. Together with myself two very bright and interesting colleagues from Czechoslovakia were visiting Wechsung's group: Branislav Rovan and Václav Rajlich, both working mainly in formal languages and automata theory. With them I had many interesting discussions and a pleasant excursion to Weimar. Wechsung was organizing a great scientific event in 1979: a conference in honor of Gottlob Frege. One hundred years before, in 1879, Frege, a

---

[16] Actually the tower had been built in 1970 to host the research center of Carl Zeiss (the famous manufacturer of cameras and optical systems), but had been donated to the university, which remained there until 1995.

professor in Jena, published his first book: *Begriffsschrift, eine der arithmetis-chen nachgebildete Formelsprache des reinen Denkens* (*Concept Notation: A Formula Language of Pure Thought, Modelled Upon That of Arithmetic*), a publication that marked a turning point in the history of logic, establishing a bridge between philosophy and mathematical logic. Wechsung invited me to give a talk at the conference, so I left Jena with the promise to come again for the Frege Conference next May.

My trip back to Rome was rather exciting because I was traveling by train and I had to cross East Germany. Going through the German Democratic Republic was always worrisome at that time. At the border the police used to check very carefully all people and all luggage, and you never knew what the result of the inspection was. Indeed when the train stopped at the border exactly what I would have preferred to avoid happened: a policeman took my passport and asked me to follow him to the office. I was worried that the train might leave without me but fortunately this did not happen. They wanted to know why I had been in Jena and who had invited me. I showed them the invitation letter by Gerd Wechsung and finally they took me back to the train. The next day I was back in Rome.

The beginning of 1979 was really a hectic period in my life. In January, af-ter a short visit to Prague and Bratislava where I had been invited by Józef Gruska, I was involved in an event that had important consequences for the Italian computer science world. For the first time the National Research Council made a precise step, marking the recognition that the domain of computer and system science was a well identified research field, autonomous with respect to mathematics on one side and electrical engineering on the other side: the formal creation of the National Research Group for System and Computer Engineering (GNASII). The group had its own budget for funding research projects that were carried out in universities and a coordi-nation role with respect to most CNR research centers in computer science. This development was certainly due to the actions that had been taken in the preceding years by the computer science groups (the groups GRIT and GRI that were mentioned previously), and by the control and system science community led by Antonio Ruberti. But it was also (I would say, mainly) due to two visionary academics, Giuseppe Biorci and Carlo Ciliberto, who had become Presidents of the CNR Consulting Committees for Engineering and Mathematics respectively. Being very influential members of the CNR governance Biorci and Ciliberto realized that computer science deserved to be enhanced, not only for the strategic impact of its applications, but also for its scientific value and they decided to take various actions in support of computer science research in Italy. The creation of GNASII was one of them[17].

---

[17] Another action, that will be illustrated later, had a very strong effect on the development of computer science and, also, on the growth of research activity in theoretical computer science: the promotion of a big project (the name was Progetto Finalizzato Informatica).

Eight years had passed since the Italian computer scientists had turned down the offer from the system science community to gather in a unique group. The fact that the alliance between the two communities of computer and system science was finally being shaped, and was no longer felt by computer scientists to be a dangerous step, was due to the numeric and cultural growth of the computer science community in Italy. It was very clear in the Italian scientific world and in the CNR milieu that informatics had emerged, day by day, as the main new field in the development of engineering. In a meeting held in Rome in January 1979, Luigi Mariani, a renowned professor of system science and bioengineering, was elected president of GNASII. At the same time I was elected director. The group was organized into 20 local units based in the major universities, and in 13 thematic subgroups (computer systems, software engineering, performance evaluation, databases, theoretical computer science and artificial intelligence, image and signal processing, bioengineering, system and control theory, optimization, system modeling, plus three subgroups devoted to applications in economics, the environment and industrial automation).

The first report of GNASII offers a view of what the Italian groups in theoretical computer science were working on in 1979. First of all it is interesting to observe that theoretical computer science and artificial intelligence were still considered strongly related topics. In fact the research carried out in the theory subgroup was classified under six titles, four of which (knowledge representation, problem solving, natural language understanding and robotics) today would be clearly considered artificial intelligence topics. The other two topics, foundations of programming and computational complexity, could instead be deservedly considered theoretical computer science topics. In the first of these two domains a primary role was played by the groups in Pisa (where Ugo Montanari and his younger colleagues, Alberto Martelli and Rocco De Nicola, were working on the semantics of cooperating sequential processes), in Torino (where denotational semantics and λ-calculus were being studied by Mariangiola Dezani and Mario Coppo) and in Rome (where Alberto Pettorossi was gathering the results of his PhD in Edinburgh on program annotation and transformation). Concerning computational complexity, the main activity reported was the work carried out in Rome about structure preserving reductions and about the classification of optimization problems according to approximability.

At the beginning the direction of GNASII required a considerable effort both in terms of organizing the evaluation of research projects and of interacting with CNR bureaucracy. But my main problem at that time was taking care of myself. In the same month I separated from my wife and I had to solve the usual elementary problems like finding an apartment and reorganizing my daily life from scratch. My colleague and friend Michele Missikoff was kind enough to invite me to share his apartment and this was already a good step forward. Then after a couple of months I met the charming woman who would become my second wife, Anna, and soon the light started to appear again.

At the beginning of May I returned to Jena for the *Frege Konferenz*. The conference was organized by the *Sektion marxistisch-leninistische Philosophie* (sic!) of the University of Jena. Most of the papers were in German and in Russian, and of course the majority of them were devoted to aspects of mathematical logic and of pure logic, but thanks to Gerd Wechsung the conference had also been open to topics in theoretical computer science. Among the invited speakers Peter van Emde Boas gave a talk about program semantics, Burkhard Monien about connections between linear bounded automata and the knapsack problem, and Salwicki about algorithmic logic. Approximation algorithms were not yet so popular: my talk about the classification of **NP**-hard problems with respect to approximability properties was the only one addressing the issue. A visit to the paintings of Lucas Cranach, both 'the Elder' and 'the Younger', which can be seen in the region, is a wonderful memory of that event.

The sixth ICALP took place in Graz, Austria, in July 1979. Hermann Maurer was the Program Chair and the organizer. The number of papers presented in Graz was 53 and, as usual, the majority were formal languages and automata papers (21) and theory of programming papers (14). Still, a few important papers addressing algorithms and complexity issues were present. In particular Karp illustrated his recent results regarding probabilistic graph algorithms[18], Hartmanis spoke about the succinctness of language description and its implications on complexity, and Papadimitriou and Yannakakis presented results on the **NP**-completeness of constrained minimum spanning tree problems. Also some interesting papers regarding new advanced aspects of the theory of programming were Manna and Pnueli's papers on modal logics (the first work of Pnueli in this direction was only a couple of years before), and Shields and Lauer's paper on semantics for concurrent systems.

To put European theoretical research in computer science in a broader context, it may be interesting to compare the scientific profile of ICALP papers with what was happening in the US in the same year, well represented by the scientific profile of STOC papers. In May 1979, ten years after the first STOC conference, about 40 papers had been accepted at STOC; 24 of them were algorithms and complexity papers (among which were several addressing new topics that were starting to attract the interest of computer scientists, such as computational geometry, VLSI design, and parallel computing), three were formal languages papers, five belonged to the domains of logic and semantics and three concerned the theory of relational databases, a rather new domain that would gain relevance and popularity in the 1980s.

---

[18] As far as I remember, the first occasion on which Richard Karp presented his research work on probabilistic analysis of algorithms was at the Symposium on New Directions and Recent Results in Algorithms and Complexity, which was held at Carnegie-Mellon University in the spring of 1976. On that occasion another seminal paper on probabilistic algorithms was presented by Michael Rabin.

A partial explanation of the difference between research on the theory of computing in Europe and in the US was given by Grzegorz Rozenberg: "I think that the main difference between the American and the European theoretical computer science scene lies in the *heterogeneity* of Europeans. While in the US it is quite typical to have just a few *fashionable* trends (with a small number of *gurus*) followed by *everybody*, it is difficult to see how, e.g., Britons could influence French not to do formal language theory (which is beautifully developed in France) because it is not anymore fashionable in Great Britain! In my opinion this diversity is the strongest guarantee of health of theoretical computer science in Europe"[19]. My feeling is that this explanation might have been somewhat correct with respect to what happened in the 1970s, before the start of the European Union research programs, which have made European computer science research more homogeneous, although I think that, in any case, the diversity of the two scenes and in particular the greater attention that the US paid to algorithm design and complexity analysis (or, if you wish, the greater role that semantics and the theory of programs had in European research) has deeper reasons, which probably relate to the stronger development of the American computer industry and the search for efficient applications. In any case the issue may be worthy of further reflection and we leave it to future investigations on the history of theoretical computer science.

What we can underline now is the prompt reaction of the the US theory community with respect to the technological drives that, in order to be adequately addressed, required advanced and non-trivial algorithmic ideas. In the transition between the 1970s and the 1980s this was happening, for example, in the case of computational geometry (strongly related to the needs of new systems for graphics and CAD/CAM), VLSI design (Mead and Conway's book *Introduction to VLSI Systems* appeared at the beginning of 1979, soon followed by the survey paper *Let's Design Algorithms for VLSI Systems* by Hsiang-Tsung Kung), and parallel computing (a subject that had started to be addressed since the first parallel computers, such as ILLIAC IV and CDC STAR-100, had been built and that is still an active research field today in relation to the modern multi-core architectures).

Going back to ICALP, as usual, in Graz we held a meeting of the EATCS council and a general assembly. Mike Paterson resigned as EATCS President and Arto Salomaa became the new president. Following Ron Book, another computer scientist from North America, Derick Wood[20], at that time a pro-

---

[19]   *The Magic of Theory and the Theory of Magic* in *People & Ideas in Theoretical Computer Science*, C.S. Calude, Ed., Springer, 1998.

[20] Derick Wood passed away in 2010 at the age of 70. He was born in Great Britain and was a very prolific scientist. He devoted himself to the study of formal languages (both Chomsky languages and Lindenmayer languages) and to a variety of other subjects, including data structures and computational geometry. After being a professor at McMaster and at Waterloo he moved to Hong Kong, to the University of Science and Technology.

fessor at McMaster University in Canada, was elected as a member of the council.

In the meantime the organization of the summer school that we had planned to hold in Udine was proceeding. The title of the school was *Analysis and Design of Algorithms in Combinatorial Optimization*. As we have seen, since the publication of the first papers presenting approximation algorithms for hard optimization problems, the subject had raised a lot of interest in the computer science community, but more recently the mathematicians working in operations research had also started to follow the research results in this field. In the last few years the complexity analysis of optimization problems and the design of approximation algorithms had indeed become very hot, and both communities were rather excited.

The interplay between computer science and operations research had started to raise some appeal in Europe since 1976, when Jan Karel Lenstra, Alexander Rinnooy Kan and Peter van Emde Boas organized a *Symposium on Interfaces Between Computer Science and Operations Research* in Amsterdam at CWI. The proceedings were edited by the same organizers and appeared two years later, in 1978. In the introduction the editors say: "The Symposium grew out of the observation that the disciplines of computer science and operations research, though never far apart, seem to exhibit more and more interaction and that many interesting developments in both areas occur at the interfaces between them. On one hand, the theory of operations research has found large scale applications only through the power of modern computing devices, and its development has benefited from the study of problem complexity and the design and analysis of algorithms. On the other hand, problems arising in the design and analysis of operating systems have led to an increased interest in queueing and scheduling theory. Thus, computer science contributes to the solution of operations research problems and asks for the solution of such problems at the same time. The proceedings demonstrate the importance of both aspects".

At the CWI Symposium many outstanding scientists had taken part. Among them Walter Savitch, Richard Karp and Peter van Emde Boas[21] who could be classified as computer scientists, and Eugene Lawler (a professor in Berkeley), Jan Karel Lenstra and Alexander Rinnooy Kan[22] were more familiar to the operations research community. The symposium also had the

---

[21] Peter van Emde Boas, born in 1945 is a renowned Dutch computer scientist. He made research contributions in a wide variety of fields, from abstract computational complexity to analysis and design of algorithms and data structures, and program semantics and logics. He is rather famous for having invented a particularly efficient data structure, known as the 'van Emde Boas priority queue', in 1975.

[22] At that time Jan Karel Lenstra (born in 1947) and Alexander Rinnooy Kan (born in 1949) were, respectively, at CWI and at Erasmus University Rotterdam. Later Jan Karel would continue his career as mathematician and became Director of CWI from 2003 to 2011, while Alexander would become a very influential business leader and in 2015 was elected to the Dutch Senate. It has to be noted that two brothers of Jan Karel, Arjen and Hendrick Willem, are also outstanding mathematicians, both specializing in computational

merit to present new approaches that were still not so well known. In his survey on machine models, after introducing Turing machines and Random Access Machines (RAMs), Walter Savitch introduced the P-RAM, that is the parallel version of the RAM, inspired by the first examples of parallel machines that were already available. As we will see, throughout the 1980s the P-RAM and other, more realistic, parallel machine models would stimulate a great amount of research work. Another subject that was introduced at the symposium by Richard Karp was probabilistic approximation algorithms for the Euclidean traveling salesman problem, a paper that became quite famous and now has reached more than 500 citations.

In my view the relevance of the symposium and of the great interest that it aroused in both the communities of OR and CS is deeper than is claimed in the preface of the proceedings. At that time the operations research community was not yet fully aware of the results that had been obtained by computer scientists in the fields of computational complexity and approximation algorithms for optimization problems. Most OR people continued with the old approach of publishing an algorithm together with tables of running times required by the algorithm on a given set of inputs, with almost no interest for complexity results proving the **NP**-completeness of the given problem, or for the possible existence of algorithms able to compute approximate solutions with a guaranteed approximation ratio. Frequently the only aim of their scientific work was to provide a new heuristic and show that in 'most cases' this heuristic gave 'good' results, without having any precise idea of what 'most cases' and 'good' meant. On the other hand, many computer scientists seemed to ignore that the problems (especially the optimization problems) that they were studying with the aim of assessing their complexity and their approximability properties had indeed been studied by OR people for decades, frequently with techniques and results that definitely deserved being better known and better analyzed[23].

Another remarkable event took place in 1976, the *Symposium on New Directions and Recent Results in Algorithms and Complexity* held in April at Carnegie-Mellon University. The symposium had been organized by Joe Traub and the Program Committee consisted of Allan Borodin, Richard Karp, Donald Knuth and Joe Traub. All important research directions in the field of algorithm design and analysis, and of problem complexity, were addressed. The first invited lecture was given by David Johnson and regarded *Approximation Algorithms for Combinatorial Problems: Prospects and Limitations* and contained an outline of the field, identifying hard optimiza-

---

number theory. With László Lovász they designed the first polynomial time algorithm for factoring polynomials with rational coefficients.

[23] With the later development of the studies on approximation algorithms, and in particular in the 1990s with the discovery of very effective approximation algorithms in which combinatorial arguments were cleverly mixed with classical OR techniques (linear programming, semi-definite programming, primal-dual algorithms etc.), people in OR and in CS have developed a common background of expertise.

tion problems that could be approximated in polynomial time with a constant approximation ratio, and problems that allowed the construction of a polynomial-time approximation scheme. Among the other talks Michael Shamos and Franco Preparata spoke about computational geometry, Michael Rabin about probabilistic analysis of nearest pair algorithms, Richard Karp about probabilistic analysis of heuristic search methods, H.T. Kung and Joe Traub about parallel algorithms, Donald Knuth about the generation of numerical distributions, and Mike Paterson about the analysis of hashing algorithms.

One of the papers presented at the symposium created a lot of rumor in the community whose scientific interest was at the border between operations research and computer science, and concerned an interesting new approximation algorithm for one of the classic, paradigmatic combinatorial optimization problems: the traveling salesman problem. In the paper *Worst Case Analysis of a New Heuristic for the Traveling Salesman Problem* Nicos Christofides presented the first algorithm that was able to guarantee an approximation ratio of 3/2, or, in other words, a maximum relative error of 50% for those instances that satisfy the triangular inequality (the previous best approximation algorithm for the same set of instances could only guarantee a relative error of 100%). While no approximation algorithm was known for general instances of that problem (it had already been proved that no approximation algorithm could exist for this case unless **P=NP**), the discovery of a 'good' approximation algorithm for metric instances was in any case a very important result. Even more important, after forty years Christofides' algorithm has not yet been beaten. In Italy the result was widely publicized when, in May 1977, Nicos Christofides and Paolo Toth organized a summer school on Combinatorial Optimization in Urbino in which various Italian and foreign experts in graph theory and operations research took part, such as Silvano Martello, Peter Hammer, Jan Karel Lenstra, and Alexander Rinnooy Kan. The proceedings of the school, edited by Christofides, Mingozzi and Toth, were published in 1979 and contained a survey on the traveling salesman problem written by Christofides[24].

The summer school that we wanted to hold in Udine was also at the border between the two domains of computer science and operations research. For this reason I decided to organize the school jointly with my colleague Mario Lucertini, a professor of operations research. But while we were working on the organization, various events in the field of algorithms and complexity marked the first months of 1979.

The first great event that in 1979 shocked the world of combinatorial optimization had been the discovery that linear programming problems could be solved in polynomial time. The linear programming problem consists of

---

[24] For a full account of the history of the traveling salesman problem and of related problems up to the late 1980s, see *The Traveling Salesman Problem. A Guided Tour of Combinatorial Optimization* by E. L. Lawler, J. K. Lenstra, A. H. G. Rinnooy Kan, and D. B. Shmoys.

finding the optimum (minimum or maximum) value of a linear function in a multidimensional space bounded by linear constraints, a problem that has a myriad practical applications. Since Cook and Karp had defined the notion of **NP**-completeness a big unresolved question was about the complexity of linear programming. The best known algorithm for the solution of linear programming problems was the *simplex method* (invented in the 1940s by George Dantzig) but this method, although in most cases very efficient, had been shown by Klee and Minty in 1972 to require exponential time for some worst case instances. On the other hand nobody had been able to prove that (the decision version of) linear programming was an **NP**-complete problem. In a sense the combinatorial structure of linear programming is not 'rich' as an **NP**-complete problem should be. The discovery of a polynomial time algorithm for linear programming came then unexpectedly, out of the blue. The discovery was due to Leonid Khachiyan, a 27 year old Soviet mathematician[25]. His work was presented at the International Mathematical Programming Symposium in Montreal, a few months after publication in the Russian journal Doklady. His approach, called the *ellipsoid method*, was not practical, due to the high exponent of the polynomial, but it was a real breakthrough for the world of operations research and computer science, since it proved that the design of polynomial time algorithms for linear programming was possible and in fact opened the way to other, more practical, algorithms that were designed in the following years[26]. Although solvable in polynomial time, in 1979 linear programming was shown to be **P**-complete (i.e. it was shown to be among the hardest problems in the class **P**) by Dobkin, Lipton and Reiss.

Also during 1979 was the publication of a fundamental result concerning graph algorithms. In fact in that year Richard Lipton and Robert Tarjan published the paper *A Separator Theorem for Planar Graphs* that had a very strong impact. In the paper, which now counts more than 1300 citations, it was shown how it is possible to divide a planar graph into two balanced components by eliminating a small number of vertices. The result, which allows us to apply the divide and conquer technique in algorithms for planar graphs, found a very large series of applications ranging from linear time algorithms for single source shortest paths to variations of Gaussian elimination for sparse symmetric systems of linear equations.

---

[25] Leonid Khachiyan, of Armenian origin, was born in 1952 in Leningrad (St. Petersburg). His 1979 paper earned him, in 1982, the prestigious Fulkerson Prize of the Mathematical Programming Society. In 1990 he became professor at Rutgers University. He died in the US at the age of 52. "He was among the world's most famous computer scientists", said Haym Hirsh, chair of the computer science department at Rutgers. Khachiyan's fame began in 1979 when Craig Whitney in The New York Times called him: "the mystery author of a new mathematical theorem that has rocked the world of computer analysis" (http://www.cs.rutgers.edu/Khachiyan/).

[26] In 1984 Narendra Karmarkar, a student of Richard Karp, invented a more efficient algorithm, based on the so-called *interior method*, suitable to be used in practice.

Another major breakthrough that appeared in 1979 was the identification by Nick Pippenger of the class of problems that can be solved in poly-logarithmic time with a polynomial amount of 'hardware' (i.e. a polynomial amount of processors in a parallel machine or a polynomial amount of gates in a circuit). The paper *On Simultaneous Resource Bounds* appeared at FOCS 1979 and the importance of the class, a subset of the class **P** of problems that can be solved in polynomial time, was immediately recognized. The class was later called **NC**, Nick's Class, in recognition of the scientist who had contributed to its definition and to the study of its properties. Among the relevant aspects of the class was the fact that its existence also had important implications for **P**-complete problems, that is those problems in **P** that are so structurally rich, that all polynomially solvable problems can be reduced to them by means of reductions that make use of a logarithmic amount of work space[27]. In fact, since **P**-complete problems are believed to be outside of **NC**, then such problems can be identified as those polynomially solvable problems that are inherently sequential, that is their solution cannot be substantially accelerated by using a parallel computer.

But probably the most remarkable event in 1979 that affected the fields of the theory of computing and of operations research was the publication by Mike Garey and David Johnson of their book on **NP**-completeness: *Computers and Intractability: A Guide to the Theory of **NP**-Completeness*. David Johnson describes how the book came out as follows[28]: "In April 1976, Mike and I attended a conference at Carnegie-Mellon University on *New Directions and Recent Results in Algorithms and Complexity*, where I gave a talk on the various types of approximation guarantees we had seen so far. Afterwards, at a coffee break, an editor for the Prentice Hall publishing company came up to me and suggested that Mike and I write a book on approximation algorithms. In thinking about that proposal, we realized that what was needed, before any book on approximation algorithms, was a book on **NP**-completeness, and by the time we left the conference we were well on our way to deciding to write that book ourselves".

Although the book was indeed a book on **NP**-completeness, in a sense it marked a transition between the early research work on **NP**-completeness of decision problems and the new era of study of approximation algorithms for hard optimization problems. In fact the book, a masterpiece that has collected 97,000 citations to date, contained a detailed presentation of the theoretical basis of **NP**-completeness and a list of 320 **NP**-complete problems (plus a list

---

[27] By 1979 various examples of **P**-complete problems had already been identified, the first one discovered by Cook in 1974: the so-called *path system accessibility*. In 1979 the above mentioned linear programming problem was also proved to be **P**-complete.

[28] David S. Johnson, *A Brief History of **NP**-Completeness*, *https://www.math.uni-bielefeld.de/documenta/vol-ismp/50_johnson-david.pdf*.

of problems whose complexity was unknown[29]) but it also presented concepts and results that formed the first body of knowledge on approximation. In particular the chapter on approximation contained the notions of **NP**-*hard problems* (optimization problems whose solution in polynomial time would imply the solution in polynomial time of **NP**-complete decision problems), of *strong* **NP**-*hard problems* (problems that are **NP**-hard even when restricted to instances in which the integers that appear in the input are bounded by a polynomial function of the input size) and of *pseudopolynomial problems* (problems for which an algorithm exists that finds the optimum solution in an amount of time which is polynomial in terms of both the input size and the value of the maximum integer appearing in the input instance), along with all results concerning approximability that could be implied for such classes of problems. The majority of this material had appeared the year before in the paper *Strong* **NP**-*completeness Results: Motivation, Examples and Implications* jointly written by David Johnson and Mike Garey. A winning idea we had in the organization of the school that year was to include a copy of the book in the registration fee, so that all participants found a brand new copy of the 'Bible' of **NP**-completeness in their bags when they registered.

The Udine School then came in a period in which the interest in the topics that we wanted to address was very lively. Beside inviting some of the scientists who had taken part in the CWI Symposium, Richard Karp, Eugene Lawler, Jan Karel Lenstra, and Alexander Rinnooy Kan, we decided to invite the scientists who had published the first important results concerning approximation algorithms and approximability properties of optimization problems, David Johnson and Azaria Paz. Finally we also invited some of our Italian colleagues actively working in the field of algorithms and data structures (Fabrizio Luccio) and in operations research (Giorgio Gallo and Francesco Maffioli[30]).

The program of the school was organized in three streams, the three major research themes that we envisaged in the field. The first theme concerned algorithmic techniques for solving specific optimization problems: variations of the minimum spanning tree problem were presented by Maffioli, a survey on approximation algorithms for the bin packing problem was presented by Johnson, and algorithms for a generalization of the network flow problem were presented by Lawler. The second theme regarded general algorithms and data structures for optimization problems and was addressed by Lenstra and Rinnooy Kan, who spoke about recursive approaches for enumerative methods, and by Luccio, who presented advanced data structures, in particular for

---

[29] Note that the list still included 'linear programming'. For most of the problems in the list we now have either the proof that they are **NP**-complete or the proof that they are in **P**, with the only relevant exception of graph isomorphism.

[30] Francesco Maffioli was one of the leading Italian figures in the field of combinatorial optimization and operations research. His contributions were particularly relevant in the areas of telecommunications, network design, and network algorithms. He passed away in 2012 at the age of 71.

bidimensional memories. Finally, the third theme, the subject of the school that was definitely the most interesting for us, concerned the complexity of combinatorial optimization problems and approaches for solving hard problems. Along this line Johnson and Paz spoke about approximation algorithms and characterization of approximable and non-approximable problems, while Karp spoke about probabilistic algorithms. I spoke about reducibility among optimization problems.

The school lasted two weeks and was rather successful. Thanks to the outstanding level of the lectures, to the perfect support of CISM (and of our colleague Paolo Serafini in particular) and with the complicity of Udine's 'spirits', the participants, young researchers from the US, Europe and Israel (most of them now full professors), were extremely happy. Many fruitful collaborations and friendly relations began that year. In particular, in a memorable dinner that I had with Jan Karel Lenstra (Fig. 8.8) and Alexander Rinnooy Kan the scientific 'marriage' between my student Alberto Marchetti Spaccamela and their student Leen Stougie was organized, a partnership that is still alive and productive today. The agreement was finalized with Friuli's excellent *Picolit* dessert wine. It would be difficult to list all the young attendees who later had an important role in the development of theoretical computer science. Just as examples I remember: John Franco, now a professor in Cincinnati, Titti Guerra, a professor at Georgia Tech, Elena Lodi, a professor in Siena, Bill McColl, later a professor at Warwick and Oxford, Linda Pagli, a professor in Pisa, and Uri Zwick, a professor in Tel Aviv. With the attendees from the Polytechnic University of Catalunya, Jaume Barceló and Xavier Berenguer, we recognized the existence of a deep common interest in the research field of algorithms for combinatorics and for optimization problems, which one year later led our Catalan friends and their colleague Josep Díaz to organize a follow-up of the Udine school in Barcelona, the *School on Combinatorics and Complexity of Algorithms*. This was also the beginning of a strong scientific relationship and of a warm long-lasting friendship between our group in Rome and the Barcelona group.

Participation in the Udine school and discussions with David Johnson and Azaria Paz about approximation encouraged us to continue the study of conditions for the existence of approximation algorithms. In particular, for a while we concentrated on the relationship that existed between the charac-

Fig. 8.8 Jan Karel Lenstra

terizations of approximable problems provided by David Johnson, based on
the notions of strong **NP**-completeness and pseudopolynomiality, and the
necessary and sufficient conditions for approximability studied by Paz and
Moran. We also, for the first time, by means of a suitable modification of
Cook's theorem, proved the existence of a problem (the weighted maximum
satisfiability problem) that was complete for the class of **NP** optimization
problems, that is, for those problems whose corresponding decision problems
belong to **NP**. On this subject I gave a talk at Helena Rasiowa's seminar
when, in November, I was invited by Andrzej Salwicki to visit his institute
in Warsaw.

A year later, two events confirmed the widespread interest in the subject of
complexity in combinatorial optimization, and in the issue of approximability,
in the community at the frontier between computer science and operations
research.

The first event was the European conference on Operations Research
(EURO IV) which took place in Cambridge, England, at the end of July
1980. Approximation algorithms were one of the major themes addressed at
the conference and one session was devoted to the characterization of approximable problems. A few groups working in this domain were present and for
us it was a good chance to compare the different approaches. In particular
great interest was raised by the presentation of Bernhard Korte (Fig. 8.9).
He was the director of the Research Institute for Discrete Mathematics of
Bonn University and was one of the most prominent leaders in operations
research in Europe[31]. At the conference he presented a paper with Rainer
Schrader on necessary and sufficient conditions for the existence of approximation schemes that is still a landmark in the field. He also gave an overview
of the field in his classic, overwhelming style that aroused enthusiasm in the
audience.

Fig. 8.9 Bernhard Korte

---

[31] Bernhard Korte (born in 1938) joined the faculty of Bonn University in 1972. Among his
former students are some outstanding figures, such as Achim Bachem, Martin Grötschel,
Rainer Schrader, and Gerhard Reinelt. More recently, an important contribution of Bernhard Korte is the book *Combinatorial Optimization: Theory and Algorithms*, written in
cooperation with Jens Vygen and published in 2008. Korte's work is not only of a theoretical nature but also concerns applications of operations research, in particular in the field
of chip design.

Many Italian experts in operations research were at the conference. Two young researchers had come with me from Rome, Claudio Leporelli and Paola Bertolazzi, former students of Mario Lucertini and Sergio De Julio (rather gifted, I would say, as both in the future became very good scientists and moreover Claudio became chairman of our Department and Paola director of the IASI institute of the Italian National Research Council). At the conference Francesco Maffioli was also present; together with him, Claudio, and Paola we had an unforgettable excursion to Ely Cathedral.

The second event was the school on *Combinatorics and Complexity of Algorithms*, organized by Berenguer and Díaz, that took place in Barcelona in September 1980. The school was attended by 65 young researchers, 30 of whom were Spanish and Italian (including our group from Rome). A report on the school written by Alexander Rinnooy Kan appeared in the journal *Qüestiió*, where most of the papers presented at the school were later published. In this report it is stressed that: "whereas many lectures in Udine were of an introductory nature, most of the Barcelona lectures dealt with recent and rather advanced results. If the participants were overwhelmed by these, that was certainly not apparent at the conference dinners and its aftermath, where various eloquent toasts were proposed to further summer schools in this area." Again most lecturers were top level; beside classical topics in operations research and combinatorial optimization presented by some of the speakers (e.g. Nicos Christofides at that time at Imperial College and Alexander Rinnooy Kan from Erasmus University Rotterdam), in other lectures stimulating new issues in algorithm design and complexity theory were addressed. Again in Rinnooy Kan's report we can read that: "L. G. Valiant of the University of Edinburgh discussed his very recent work on a schema for fast parallel communication and talked about the complexity of counting problems[32] ... R. M. Karp of the University of California at Berkeley discussed his probabilistic analysis of connectivity and matching algorithms ... P. Flajolet of INRIA gave a survey of his recent work on the analysis of dynamic data structures."

The topics presented at the school by these outstanding scientists were indeed very recent, remarkable and original contributions. In particular, in 1979 Les Valiant (Fig. 8.10) had published two very important papers concerning

---

[32] Les Valiant, born in Budapest in 1949, obtained the Turing Award in 2010 with the following citation: "For transformative contributions to the theory of computation, including the theory of probably approximately correct (PAC) learning, the complexity of enumeration and of algebraic computation, and the theory of parallel and distributed computing". Parts of these topics were addressed in the lectures given by Valiant in Barcelona. In a paper in the Communications of ACM, where the Turing Award is announced, the following remark is reported regarding Valiant's approach to research: "He looks at what many people would see as very complex problems, and he breaks them down into something beautiful and elegant, really getting at the heart of the problem," says Jennifer Chayes, managing director of Microsoft Research New England. "He thinks with such amazing clarity. Part of it is his solution, but a lot of it is his definition and articulation of these problems."

Fig. 8.10 Leslie Valiant

the complexity of enumeration problems, i.e. the problems that consist in counting the number of solutions of a combinatorial problem: *The Complexity of Computing the Permanent* and *The Complexity of Enumeration and Reliability Problems*.

In these papers he considers polynomial time solvable counting problems (e.g. counting the number of spanning trees in a graph) and he defines the complexity class $\#P$, that is the class of counting problems that are as hard as the problem of determining the number of accepting computations of a polynomial-time non-deterministic machine and, henceforth, are at least as hard as the **NP**-complete problems (e.g. computing the permanent of a $(0, 1)$ matrix or computing the number of perfect matchings in a bipartite graph). These papers (that now count more than 2000 and more than 1500 citations respectively) and a few other minor contributions on the same subject opened a new research domain and can definitely be considered among the major scientific contributions that appeared in 1979. In the same year Valiant presented the paper *Completeness Classes in Algebra* at STOC, again a paper of a foundational nature, which established the ground for an algebraic approach to computational complexity. Also the work presented by Philippe Flajolet has a place in the history of theoretical computer science. In the mid-1970s Philippe had moved his research activity toward the combinatorial analysis of data structures and between 1979 and 1980 he published a series of papers (some of them together with Jean Vuillemin, who also had shifted his scientific interest from semantics to the theory of algorithms) in which, making use of rather sophisticated mathematical techniques, the average case performance of dictionaries and other dynamic data structures, subject to an arbitrary sequence of operations (e.g. insert, delete and search), was analyzed.

Finally, with great disappointment, Alexander Rinnooy Kan notes in his report that one of the invited lecturers, Jaroslav Nešetřil (Fig. 8.11) of Charles

Fig. 8.11 Jarik Nešetřil (right) with Giorgio Ausiello

University (Prague)[33], who was already known as an excellent scientist, "was refused permission to attend the school at the last instant" and could not come. It was 1980, nine more years and finally the 'wall' would come down.

---

[33] Jaroslav (Jarik) Nešetřil, born in 1946, is an extremely prolific scientist working in the fields of discrete mathematics and combinatorics. Two of his books in particular have become very popular: *Invitation to Discrete Mathematics* written in collaboration with J. Matoušek, and *Sparsity – Graphs, Structures, and Algorithms*, written with P. Ossona de Mendez. For his research activity Jarik has received a long series of recognitions and awards. He is also well known for being a very talented artist.

Chapter 9
Relations

After the Udine School, Marco Protasi, Alberto Marchetti Spaccamela and I were very excited and started to conceive the publication of two books on algorithms: one devoted to fundamental algorithms and one devoted to combinatorial optimization problems and approximation algorithms. The first one was completed and published in Italian four years later. The writing of the second one was more difficult because the field was exponentially growing throughout the 1980s and even more dramatically in the 1990s. Any time we thought we had a suitable draft a new breakthrough result came to change the picture. Finally after twenty years, three drafts and three more authors (Pierluigi Crescenzi, Giorgio Gambosi and Viggo Kann) the book appeared in 1999.

But the field of theoretical computer science was rapidly evolving and new domains were being addressed at the end of the 1970s and at the beginning of the 1980s in the international community. Interesting contributions came also from Italian computer scientists, also thanks to various actions that Italian public institutions undertook in those years for the promotion of informatics.

In 1979 the Italian National Research Council launched a very important initiative: the *Progetto Finalizzato Informatica*. The idea to promote a series of *Progetti Finalizzati* (multiannual targeted projects) had been considered since 1972 when Alessandro Faedo, a mathematician from Pisa that we have already mentioned for his very effective role, had become President of CNR, with the aim of establishing, in various strategic domains of science, a stronger connection between basic research, industrial development and applications. The funds for these projects should have come directly from the government, outside of the ordinary annual CNR budget, while CNR was in charge, initially, to conduct the feasibility study and then, in case the projects were approved and funded by the government, to organize and manage their execution. When in 1976 the new President of CNR, Ernesto Quagliarello, was appointed and Biorci and Ciliberto became Presidents of the two CNR Consulting Committees for Engineering and Mathematics, respectively, they decided, as we already said, to put in place a series of actions for the pro-

© Springer International Publishing AG, part of Springer Nature 2018
G. Ausiello, *The Making of a New Science*,
https://doi.org/10.1007/978-3-319-62680-2_9

motion of informatics, among these the execution of a feasibility study for a
Progetto Finalizzato Informatica (PFI). The project was eventually approved
by the government in January 1979 with a budget that expressed in today's
currency would correspond to about 500 million euros over 5 years.

The kick-off of PFI came in June 1979. The project was organized into
three subprojects, essentially devoted to computer architectures and networks, databases and applications in public administration, and computer
applications in industry. Director of the project was Angelo Raffaele Meo,
professor of computer engineering at the Polytechnic of Turin, and directors of the three subprojects were Ugo Montanari, professor in Pisa, Paolo
Bronzoni, researcher of CNR in Pisa, and Riccardo Zoppoli, professor of system science in Genoa. Such choices (in particular the appointment of Ugo
Montanari) were indeed a recognition of the relevant role that basic research
should play in the project (although, clearly, also practical applications were
expected in all subprojects). Also the Scientific Council of PFI, in which I
was involved, had a substantial presence of theoreticians. Besides, the fact
that the project concerning computer applications in industry was entrusted
to a professor of system science was another sign that CNR recognized the
value of the cooperation between computer scientists and system scientists,
again following the French example, where the Research Institute for Informatics and Automatics, IRIA, had just become INRIA in recognition of
a new national role that the institution had gained, and where INRIA new
President Jacques-Louis Lions[1], appointed in 1979, was a professor of applied
mathematics and control theory.

In a series of meetings that were held in Milan and in Pisa a fierce debate
in the Italian database community took place on what should be the main
targets of the research conducted inside the subproject devoted to databases
and applications in public administration. On one side stood the university groups interested in developing theoretical and technological aspects of
distributed databases; on the other side the groups interested in methodologies and tools for database design. Other groups instead were interested
in applications and in particular wanted to develop applications for local
administrations. Finally, it was decided to split the subproject into various
targets: 4 application-oriented targets (essentially information systems for
labour offices, for local public administrations and for handling geographi-
cal information) and 2 targets oriented toward new models for distributed
computing (Objective DATANET) and methods and tools for conceptual relational database design (Objective DATAID). Most of the university and
CNR researchers participated in these two projects. I decided that part of
our group should participate in the DATAID project because for two or three

---

[1] Jacques-Louis Lions (1928–2001) was an outstanding mathematician, specialized in partial differential equations and stochastic control, and a very influential scientist. After
being President of INRIA he was also appointed director of the Centre National d'Etudes
Spatiales and elected President of the International Mathematical Union and subsequently
President of the French Academy of Sciences.

years, together with my younger colleagues Carlo Batini, Sandro D'Atri and Marina Moscarini, we had started to look at some problems concerning relational database design and we envisaged a great room for interesting theoretical work in this domain.

As we mentioned before, the relational database model had been introduced by Ted Codd in 1969 in an IBM technical report followed by the 1970 seminal paper: *A Relational Model of Data for Large Shared Data Banks* appeared in the Communications of the ACM. This paper, which now has collected more than 9600 citations, had a revolutionary impact in the field of databases, for several reasons. Until that moment, such as for example in the IBM Information Management System (IMS), large collections of data stored in secondary memory were essentially represented by providing an explicit description of the record structures that were contained in the files; also, the relations between data were represented in terms of links between physical records, possibly organized in a hierarchical way. In a sense, therefore, there was no clear distinction between the logical and the physical data organization.

In the relational approach, instead, the logical model seen by the user consisted of a set of relations, each one corresponding to a list of attributes (e.g. the relation EMPLOYEE might consist of the list of attributes (CODE#, NAME, AGE)) and the database would consist of a set of tables, each one corresponding to a relation and each one consisting of a set of tuples defined over the domains of the attributes (e.g. all employees stored in the table). Relationships among data can be themselves represented as tables. In this way, for the first time, the relational model allowed us to realize the independence between the logical and the physical level in the database system. The advantages of the model were numerous. First of all, exploiting the mathematical properties of relations, it was possible to introduce formal and precise languages both for defining the scheme of the database at the logical level and for querying the database. Besides it was possible to provide, again in a formal and precise way, most of the constraints that the data instances had to satisfy (such as, for example, the fact that there should be a bijection between code number and name of an employee or that the age of an employee should be greater than or equal to 18). As we said before, the release of commercial versions of relational database management systems started only at the beginning of the 1980s but, after the publication of Codd's paper, the interest in the relational model grew immediately in the theory community and prompted a large series of studies mainly concerning the logical constraints that should have been satisfied by relational databases and the computational (expressive) power of query languages.

In the obituary dedicated to Edgar Codd, Chris Date, one of the most important scientists in the field of databases says[2]: "Edgar Codd, ... single-handedly put the field of database management on a solid scientific footing.

---

[2] C. J. Date, *Edgar F. Codd 1923–2003*, in *Memorial Tributes*, National Academy of Engineering, Volume 12, The National Academies Press (2008)

The entire relational database industry, now worth many billions of dollars a year, owes its existence to Ted's original work; the same is true of all relational database research and teaching programs in universities and similar organizations worldwide. ... The relational model is widely recognized as one of the great technical innovations of the twentieth century. Ted described it and explored its implications in a series of staggeringly original research papers published between 1969 and 1981. The effect of those papers was twofold: First, they changed for good the way the IT world perceived the database management problem; second, they laid the foundation for a whole new industry".

The first important contributions concerning the relational model had been those regarding normal forms and data dependencies, in particular functional dependencies, that is the relationship between those attributes that form the key of a relation and the attributes that functionally depend on this key. In the original paper by Codd in 1970 only the first normal form was foreseen, requiring that the domains of all attributes only consisted of atomic values, but soon, between 1972 and 1977, Ted Codd, Raymond Boyce and Ron Fagin introduced other normal forms aimed at eliminating data redundancy, avoiding various kinds of anomalies and reducing the need for restructuring the tables upon extensions of the database. The work of William Armstrong, presented, as we already mentioned, at the 1974 IFIP World Congress, was devoted to data dependencies. Armstrong introduced axioms for data dependencies that allow us to derive all dependencies that are logically implied by a given set of functional dependencies. Also the work of David Maier (Fig. 9.1) was devoted to data dependencies; at STOC 1979 he published the paper *Minimum Covers in the Relational Database Model* containing the first results on the problem to determine a minimum set of functional dependencies that are, in a suitable sense, equivalent to the given set. A third line of research in those early days was devoted to the study of the expressive power of query languages, in particular languages based on relational algebra and languages based on predicate calculus. We already mentioned Codd's paper concerning the relational completeness of database languages that was presented at the Courant Symposium. Relational completeness, that is the fact that a query language has at least the same expressive power as relational al-

Fig. 9.1 David Maier

gebra, became the basic reference for assessing the expressive power of query languages, and using this concept Codd showed that a language based on predicate calculus such as the relational calculus is at least as powerful as relational algebra. The same issue had been analyzed by François Bancilhon at MFCS in 1978 and is one of the most studied subjects especially after the introduction of more powerful logic-based languages allowing transitivity, negation, etc. From this point of view, particularly important has been the Datalog language, essentially a subset of PROLOG used as a query language for deductive databases. The notion of deduction in databases and the use of logic languages derive from the attempt to put together logic programming and databases, started by Gallaire and Minker in 1977. Deductive databases may be queried by users not only to extract existing data but also to infer new information from existing data (for example by reporting all ancestors of a citizen from a database that contains the father-son relation). Concerning the origin of the language Datalog: "It is difficult to attribute Datalog to particular researchers because it is a restriction or extension of many previously proposed languages; the name Datalog was coined (to our knowledge) by David Maier"[3].

In Rome, at the end of the 1970s, the study of databases had taken two directions: one was devoted to the study of suitable notions of equivalence between relational database schemata and the second, closely related to the project DATAID, was devoted to conceptual database design. As we already said, the relational model had allowed us to establish a certain degree of independence between the physical organization of data and the logical schema defined in terms of relations. Although this had been an important forward step to give database users a vision of data banks that is free of implementation details and closer to their own needs, this step was not considered sufficient by database designers. In direct contact with the final user, the designer had to collect and formalize the user's requirements and identify the 'entities' that formed his view of the world (e.g., in a university environment, the entity Teacher and the entity Course) and the 'relationships' interconnecting such entities (e.g. the one-to-many relationship 'Teaches' that connects teachers and courses). This more abstract level of design, called 'conceptual design' had been introduced by Peter Chen[4] (Fig. 9.2) at the first International Conference on Very Large Databases (VLDB) that took place in US in 1975, with the paper *The Entity-Relationship Model: Toward a Unified View of Data*. Immediately, the Entity-Relationship approach became a great success (the journal version with the same title that appeared the next year in the *ACM Transactions of Database Systems* has now more than 9700 citations).

---

[3] S. Abiteboul, R. Hull, V. Vianu, *Foundations of Databases*, Addison-Wesley, 1995

[4] Peter Chen, an outstanding Taiwanese computer scientist, earned his PhD at Harvard in 1973. After spending a few years working for various US companies he became professor at UCLA and subsequently at Louisiana State. Currently he is a professor at Carnegie Mellon. The invention of the Entity Relationship model has provided a terrific contribution, both theoretical and practical, to the field of database design.

Fig. 9.2 Peter Chen

Carlo Batini and Alessandro D'Atri had started to do some interesting work in the direction of top-down methods for database design since 1978. Their idea was to describe the top-down construction of a database schema in terms of rewriting rules for graphs and hypergraphs and their paper with the title: *Rewriting Systems as a Tool for Relational Data Base Design*, had been presented in 1978 at the first *Workshop on Graph-Grammars and Their Application to Computer Science and Biology*. Of course the idea to use the same approach for conceptual design was rather natural, and one year later, in 1979, Carlo Batini presented his paper *Top-Down Design in the Entity-Relationship Model* at the first conference organized by Peter Chen on *The Entity-Relationship Approach to Systems Analysis and Design*. This explains why we thought that the research domain of conceptual design included in the DATAID project was prone to interesting theoretical developments and why Carlo Batini was, somehow naturally, asked to coordinate the DATAID project.

The DATAID project was recognized as a success story in the framework of the *Progetto Finalizzato Informatica*. The project produced not only first-class papers that appeared in top conferences and journals during the 5 years of the project, but also a methodology that was adopted by some Italian software companies[5], and various graphic tools for supporting top-down conceptual design in the entity-relationship model[6]. Six of the research groups

---

[5] Indeed when in the 1990s Carlo Batini became a member of the Italian Authority for Informatics in Public Administration the methodology (with suitable adaptations) became a standard for the delivery of software systems in reply to public tenders.

[6] In relation to the graphic systems developed in the DATAID project, it is worth underlining that the drawing of E-R diagrams had an important role in the research training of two Italian computer scientists who later gave great contributions to the field of graph

taking part in the project were universities or CNR institutes (Turin, Milan, Bologna, Pisa, Rome), five were industrial groups, including those that were, at that time, the major Italian software companies (Olivetti, ITALSIEL, Systems & Management) plus a small research center recently created in Calabria: the *Consortium for Research and Applications of Informatics* (CRAI).

It is worth devoting here a few words to CRAI because this center, created in 1978 by my colleague and friend Sergio De Julio, professor of Operations Research at the University of Calabria, by exploiting an extraordinary funding allowed by the government in the 1970s for the promotion of advanced technology research in Southern Italy, in a few years was able to achieve an important role in European database research. As a witness of such developments, I think it is worth stressing that CRAI is another success story in which, beside the ability of the management to put together researchers, developers, engineers with highly qualified skills, the role of theoretical computer science has not been negligible and has allowed this institution to develop cutting-edge research, to establish international connections with other advanced research centers and to take part in important European projects.

My collaboration with CRAI started just in June 1979 when on a flight to Cosenza, where I was invited to the Mathematics Department to talk about approximation algorithms, I had the chance to meet Sergio De Julio. Sergio had been associate professor in the Istituto di Automatica in Rome, but he left Rome shortly after I moved to the Istituto because in 1975 he became a full professor and he took a position at the University of Calabria. On the airplane we discussed research in theoretical computer science and he described to me how he wanted to develop computer research at CRAI on a solid theoretical basis although with an eye to applications. When we arrived in Cosenza I had been recruited as a consultant of CRAI with the task to start research projects in theoretical aspects of databases and information systems and to design education programs for CRAI researchers.

From October 1979 I started visiting CRAI a couple of times per month to meet regularly with some of the young CRAI researchers, in particular with Domenico (Mimmo) Saccà who was particularly brilliant and endowed with very good mathematical skills as well as with high sensitivity for practical aspects of information systems. Collaborating with us also was Alessandro D'Atri who had been a lecturer of a computer science course at the University of Calabria from 1977 till 1980. The atmosphere at CRAI was incredibly tense. All researchers were strongly motivated. They perceived that, thanks to the high profile that Sergio De Julio wanted to give to this research center, being enrolled at CRAI[7] was a great chance for their career. For the first time

---

drawing and graph visualization, Roberto Tamassia, now professor at Brown University, and Pino Di Battista, now professor at the Third University of Rome (see G. Di Battista, P. Eades, R. Tamassia, I. G. Tollis, *Graph Drawing: Algorithms for the Visualization of Graphs*, Prentice Hall, 1998).

[7] CRAI closed its activity in 1996 but thanks to the high technical and scientific level of its productive work various CRAI researchers were able to move to academia (and have now

a public initiative devoted to hi-tech was taking root in Calabria, one of the most disadvantaged regions in Southern Italy, and this opportunity could not be lost.

D'Atri, Saccà and I decided to address the study of graph and hyper-graph algorithms for the synthesis of relational database schemes and for the manipulation of functional dependencies. By the way this topic would have been perfectly suited for the DATAID project in which both CRAI and my department were participating. Our first results were presented in 1980 at the *Workshop on Graph Theoretic Concepts in Computer Science*. This was the first of a rather successful series of papers devoted to transitive closure, transitive reduction and minimal covers of functional dependencies that three years later ended up in the *Journal of the ACM* and in the *SIAM Journal on Computing*.

Besides carrying on our research work, together with Sergio De Julio we decided to organize a school or seminar devoted to the theory of relational databases in 1981. This would have been one of the first big international events devoted to this field and we wanted to bring to Italy, more precisely to Calabria, some of the top researchers with the aim, on one side, to introduce a generation of young students to the most important research topics of this new domain and, at the same time, to make the international community working in the field aware of the existence of CRAI and of its ambitious research projects.

Partly in view of the work we were carrying on in Rome on approximation algorithms and partly as a consequence of my appointment as a consultant of CRAI I decided to go for a trip to the US in the Spring of 1980. The interesting occasion was that in 1980 both the STOC conference and the SIGMOD conference, at that time, together with VLDB, one of the few database conferences, were taking place in Los Angeles, at the end of April and in mid-May respectively, and this allowed me to have some time to visit friends in Berkeley. In any case the first stop would be Urbana where my friend Franco Preparata had invited me to give a talk at the University of Illinois.

Franco has always been extremely dynamic in his research activity, always ready to tackle the most daring challenges. In that period Franco's research work was mainly devoted to two domains to which he made fundamental contributions: computational geometry and parallel computing.

Computational geometry was a somewhat recent field of research. Although since the beginning of the 1970s, in the early days of computer-assisted design and computer-assisted manufacturing (CAD/CAM), there were attempts to provide computational descriptions of geometrically complex objects, after 1975 the term computational geometry was used to denote the design of algorithms aimed at the efficient solution of geometric problems such as computing the convex hull of a set of points in an n-dimensional space, or

---

gained a good international reputation) and many others created a series of small software enterprises that are still actively operating in Calabria.

determining the correct shadows in a 3-dimensional scene under a given light source, etc. This research field was stimulated by applications in computer graphics, in mathematical visualization, in robotics, and in computer-aided engineering. Another application domain, in which geometric algorithms had a major role, such as for example algorithms for computing the intersection of polygons, was geographic information systems. Franco's first geometry papers were from 1977 (*Convex Hulls of Finite Sets of Points in Two and Three Dimensions*, appeared in the *Communications of the ACM*) and 1978 (*Triangulating a Simple Polygon*, appeared in *Information Processing Letters*, and *Finding the Intersection of Two Convex Polyedhra*, appeared in TCS). In a few years, the rich scientific production in the field lead Franco and one of his students, Michael Shamos, to publish the book *Computational Geometry – An Introduction*, a major textbook that has now collected more than 10,000 citations.

In the second domain, parallel computing, the scientific work carried out by Franco Preparata was devoted on one side to the design of specific parallel algorithms for specific parallel computing systems and on the other side to the design of new interconnection topologies. The study of parallel architectures and parallel computing models had started already in the early 1970s, but in the 1980s it reached its apex in connection with two technological developments: the design of very-large-scale integrated (VLSI) circuits, on one side, and, on the other side, the development of low cost processors that allowed us to consider the possibility of assembling massively parallel computer systems. Although the most important contributions made by Franco in the domain of parallel computing were realized in subsequent years and concerned VLSI systems, in 1980 he had already produced an important breakthrough in the field of interconnection networks. In 1979, at FOCS, in cooperation with Jean Vuillemin, Franco had published one of his landmark papers (that now has more than 1000 citations) *The Cube-Connected-Cycles: A Versatile Network for Parallel Computation* that was published in the *Communications of the ACM* in 1981. The cube-connected-cycles topology consisted of a hypercube interconnection where each vertex was replaced by a cycle so that all vertices in the network had degree three. While the advantages of a hypercube (in particular small diameter, which implies small latency) were maintained, the new interconnection scheme only required three connections per processor and had optimal characteristics for VLSI implementations.

My stay at Urbana was a great opportunity to have a close look at Franco's work and to meet with his colleagues but was also the first chance for me to meet with his family and in particular with his wife Rosamaria, a biologist, whose smile, composure and intelligence have contributed so much to build their rich scientific and cultural life. During the same period also Fabrizio Luccio (Fig. 9.3) was at Urbana as a visiting professor; at that time Fabrizio was working on magnetic bubble memories, a computer storage technology that created some interest in the 1970s and early 1980s but was soon pushed aside by new generations of very fast solid-state memory chips. Since during

Fig. 9.3 Fabrizio Luccio (center) with collaborators in 2001: among others, Linda Pagli (left), Paolo Ferragina and Roberto Grossi (right)

my stay Fabrizio and his wife Cristina were traveling, I was hosted in their house, with the only commitment to prepare pancakes at breakfast for their children, Giulio and Flaminia (now associate professor of computer science at *Ca' Foscari University* in Venice). But the most remarkable event while I was at Urbana was the telephone call that I received from Corrado Böhm in which he announced that I had succeeded in the competition for a position of full professor. Another exciting phase in my life would soon start.

A few days later I attended STOC in Los Angeles. The first impression I remember of the conference is the presence of Leonid Levin together with Lenore Blum at the registration desk. Leonid was known as the Soviet scientist who had discovered **NP**-completeness independently from Cook's work; he had just recently earned his PhD at MIT under the supervision of Albert Meyer. As usual, STOC was rich with important results. In that period, stimulated by the fundamental result of Hopcroft and Tarjan on the isomorphism of planar graphs, a certain effort was concentrated on the study of isomorphism of general graphs, one of the problems for which neither polynomial-time algorithms nor **NP**-completeness results were available. At STOC four papers were devoted to this elusive problem and in particular two of them (by Gary Miller and by Ion Filotti and Jack Mayer) concerned polynomial-time algorithms for deciding isomorphism of graphs of bounded genus (i.e. graphs embeddable on closed surfaces with a bounded number of 'holes'. Note: planar graphs have genus 0 and graphs embeddable on a torus have genus 1). Both papers (independent although related) were based on preceding work by Filotti, Miller and Reif regarding algorithms for determining the genus of a graph, whose results had been presented at FOCS 1979. Complexity and algorithm design were, as usual, the major domains addressed at STOC, but various researchers were moving to other topics. Among them, Albert Meyer had just started to study semantics and the logic of programs and presented a paper on dynamic logic, a language used to express and prove properties of programs, and Nancy Lynch, at that time at Georgia Tech, was starting her work on the theory of distributed systems, the field in which she has made very deep contributions. The theory of relational databases was addressed in

Fig. 9.4 Hartmut Ehrig (left) and Bernd Mahr

the talks of Sadri and Ullman and by Ron Fagin who spoke about the ax-
iomatization and the logic properties of special kinds of data dependencies.

At STOC I also met two good friends: Hartmut Ehrig and Bernd Mahr[8]
(Fig. 9.4). Hartmut and Bernd were working on abstract data types and
presented a paper on the complexity of data type implementations in an
algebraic context. Like me, Hartmut and Bernd were going to Berkeley after
the conference and we decided to travel together by car. Hartmut and Bernd
were not only great computer scientists but also culturally rich and the trip
was extremely pleasant. We went along Highway 1 and we stopped for the
night in Big Sur. Big Sur is still full of memories of Henry Miller who lived
there from 1944 to 1962 (our hotel had one room with Henry Miller's bath
tub!) and in this atmosphere evening drinks at the Nepenthe Bar was a ritual
ceremony.

During our trip we stopped at SRI in Menlo Park to visit Joe Goguen. The
visit was educationally very relevant for me. Joe told us about the work he
was doing in language analysis. The aim was to understand where, how and
why misunderstandings could occur in critical conversations (e.g. between
pilots and air traffic control or between a psychologist and patients under
treatment). For the first time I realized that computer science not only was
the science of computing but also had a strong epistemologic power.

The next day we were in Berkeley. With real emotion, after ten years,
I walked through Sather Gate and entered the campus. The first step, of
course, was to visit Manuel Blum and Dick Karp. In particular, I wanted to
discuss with Karp the possibility that Alberto Marchetti Spaccamela could
be accepted as a visiting researcher for one year and could do some research
work under his supervision (which indeed happened two years later).

---

[8] Hartmut Ehrig (1944–2016) and Bernd Mahr (1945–2015), both professors of computer
science at the Technical University in Berlin, were among the greatest theoreticians in
Europe and have left a profound imprint in the field of abstract data types and more
generally of algebraic approaches in the theory of programming. Their books *Fundamentals
of Algebraic Specification* (1 and 2), published in 1990 are still basic textbooks in the field.
Hartmut Ehrig also made fundamental contributions to the field of graph grammars and
graph transformations. As Ugo Montanari says in the obituary which appeared in the June
2016 issue of the EATCS Bulletin, "Hartmut has been the main originator and architect
of this area".

Fig. 9.5 Silvio Micali (left) and Shafi Goldwasser

Then I looked for Silvio Micali (Fig. 9.5), who at that time was a PhD student of Manuel. I had met Silvio[9] the year before at a school organized by the University of Lecce. Silvio had just obtained his laurea degree in Rome under the supervision of Corrado Böhm and in fact his first paper concerned λ-calculus. During the school I was struck by Silvio's intellectual energy and pleasantness. Another lecturer at the school was Shimon Even[10], who had just published his famous book on graph algorithms. Also Shimon was impressed by the brightness of this exceptional student. I think we owe it to the exuberant and overwhelming personality of Shimon, so close to the warm Sicilian character of Silvio, and to the few days that they spent together traveling through Southern Italy that Silvio decided to address his future work to algorithms and complexity and he chose to apply as a PhD student in Berkeley.

When I arrived in Berkeley Silvio was working in cooperation with another of Blum's students, Vijay Vazirani. They had just developed a new algorithm for the matching problem on general graphs that outperformed the previous algorithm that Edmonds had published twenty years before in his classic paper *Paths, Trees and Flowers*. Definitely a good start for two PhD students! But eventually Silvio's work, under the supervision of Manuel Blum and with the scientific collaboration of Shafi Goldwasser (again another outstanding Blum student), was going to be even more successful, leading Silvio and Shafi to receive the most ambitious prize: the Turing Award. Indeed Silvio's achievements had a revolutionary impact due to the introduction of unforeseen new methods in the theory of computing. In particular in the paper *The Knowledge Complexity of Interactive Proof-Systems* (presented at STOC 1985) Goldwasser, Micali and Charles Rackoff introduce the notion of

---

[9] Silvio Micali, born in 1954, has been a professor at MIT since 1983. Silvio was awarded the Turing Award in 2012 along with Shafi Goldwasser "for transformative work that laid the complexity-theoretic foundations for the science of cryptography, and in the process pioneered new methods for efficient verification of mathematical proofs in complexity theory".

[10] Shimon Even (1935–2004) was one of the greatest Israeli computer scientists. He got his PhD in Harvard and in 1974 joined the faculty at the Technion. Among his students are several outstanding theoreticians like Alon Itai, Michael Rodeh, Baruch Awerbuch and Oded Goldreich.

zero-knowledge proofs, that is an interactive proof system in which the prover can convince the verifier that she has the proof that a certain statement is true (for example that two graphs are isomorphic) without revealing any detail of the proof. The concept of zero-knowledge proofs, further developed in the paper *Proofs That Yield Nothing but Their Validity and a Methodology of Cryptographic Protocol Design* (presented at FOCS 1986 by Oded Goldreich, Silvio Micali and Avi Widgerson) and in various other papers, would lead to important consequences not only in cryptography but also in complexity theory[11]. As Avi Widgerson says in the presentation of Micali's Turing Award: "Silvio Micali is a visionary whose work has contributed to the mathematical foundations of cryptography and has advanced the theory of computation. His non-conventional thinking has fundamentally changed our understanding of basic notions such as randomness, secrets, proof, knowledge, collusion, and privacy, which have been contemplated and debated for millennia. This foundational work was a key component in the development of the computer security industry, facilitated by his patents and start-up companies. His work has also had great impact on other research areas in computer science and mathematics".

Finally it was time to have a few days of vacation. Anna had reached me in Berkeley and we left, direction Arizona. First of all we stopped in San José at the IBM Research Laboratory where I wanted to meet Mario Schkolnick, a Chilean computer scientist involved in System R, the first IBM relational database system, an ancestor of the well-known commercial IBM relational database system DB 2, in order to discuss the possibility for some of the researchers of CRAI (the research center that was being built in Calabria) to spend one year at San José under his supervision. Then we went along the classical tour: Painted Desert, Grand Canyon, Monument Valley. On the radio Bob Seeger was singing *Against the Wind*.

On May 14 we were back to Los Angeles, actually Santa Monica, for the *ACM SIGMOD Conference on Management of Data*. The series of SIGMOD conferences had started in 1975 (the same year in which the VLDB conference had been initiated) and had replaced the previous series of SIGFIDET conferences that had been going on since 1970. In 1980 the PC Chair was Peter Chen, the father of the Entity-Relationship model that was mentioned before.

The conference program shows the diversity of subjects that were addressed in that period. Most of the talks concerned relational databases although a few talks were still devoted to the network model (CODASYL) and to the hierarchical model (IMS). At the physical level the most interesting idea of those years was the possibility to realize a 'database machine',

---

[11] In the paper *Everything Provable Is Provable in Zero-Knowledge* by Ben-Or, Goldreich, Goldwasser, Hastad, Kilian, Micali, Rogaway, presented at CRYPTO 88, it is shown that zero-knowledge proofs exist for all languages that can be decided in polynomial time by means of interactive proofs (that is the class of languages IP, known to coincide with the class PSPACE of polynomial-space decidable languages).

a computer architecture oriented toward fast data management operations that should work as a back-end processor tightly coupled with an ordinary computer. On this issue were engaged, among others, François Bancilhon and Michel Scholl, at INRIA, and H.T. Kung, at Carnegie Mellon, who was studying the application to databases of his model of parallel architectures, the 'systolic' VLSI architectures, that we will illustrate below. Other important contributions at that conference, which may reflect the contemporary type of research activity, concerned physical data structures (mostly B-trees), distributed databases, various aspects of database design (from the Entity-Relationship conceptual level to the physical level) and various approaches for adding knowledge to user interfaces. Finally, in the area of database theory there were two papers (by Jeff Ullman and Douglas Stott Parker) devoted to various kinds of data dependencies.

The conference was a great chance to understand the main research trends in the field and to establish contacts in view of the school that we wanted to organize in Calabria the next year. Relational database theory appeared to me, more and more, the most interesting and promising direction to promote at CRAI. It seemed to me appealing mainly because it allowed us to reconcile application-oriented and practical issues with theoretical work, essentially the design of efficient algorithms, often graph algorithms. This was in favour of my (and De Julio's) idea that CRAI should be a center for applied research but CRAI researchers should compete at international level on cutting-edge research topics.

At the end of May I was back in Rome. First of all I had to finish my course at the university and in the meantime I also had to take care of my personal logistic situation that was still rather uneasy. Recently my colleague Marco Protasi had offered to lend me a fairly large apartment and this might allow me to better organize my new life.

In the Fall I left CNR and joined the Faculty of Engineering of Rome University. The change was quite impressive. I had been working at CNR for more than 13 years, always in scientifically restricted environments (mathematics and computer science) and the immersion in the multicultural context of a Faculty of Engineering forced me to compete with other emerging disciplines (electrical engineering, telecommunications, aeronautics, etc.) and to fight day by day to gain a scientific recognition for my work. I also soon realized that although theoretical work was highly respected, in a Faculty of Engineering you have also to show the practical relevance of your achievements and, also for this reason, I decided to give more emphasis to the work I was doing in the field of databases in the framework of the CNR project DATAID and I even started to work on a very applied project for the design of the regional health information system. I was also feeling sorry for abandoning my CNR institute. Just in that period the Research Center for Automatic Control and Computing Systems was being transformed into a permanent structure: the Institute for Systems Analysis and Informatics (IASI). As director of IASI had been nominated Lucio Bianco, a prominent figure in the field of operations research, with excellent organizational capacity[12], and I was missing a good chance to consolidate the theoretical computer science activity within the new institute, activity that was in any case being advanced by several young colleagues like Paolo Atzeni, Alfonso Miola, Michele Missikoff, Marina Moscarini, Alberto Pettorossi, Silvio Salza, all of whom eventually would become professors in various of Rome's universities.

The first few months of 1981 were a hectic time. In that period I had resumed my collaboration with CRAI and my research work with D'Atri and Saccà on functional dependencies. I was also pushing further the organization of the seminar to be held in September. Together with the local organizer, Mimmo Saccà, we found a fantastic location on the coast of Calabria, in Cetraro, at the San Michele Hotel, a very beautiful villa that had been transformed into a hotel after the war, overlooking a small private beach from the top of a cliff 80 meters high. The hotel had a huge extension of gardens and vineyards, and offered excellent sporting facilities. We also started to discuss with François Bancilhon and Nicolas Spyratos the scientific program and we began to send out invitations to some of the major figures of database theory trying to cover a wide range of topics, especially the topics that were hot in those early days of database theory research: data dependencies, acyclic database schemes, semantic models, logic models, incomplete databases and null values (Appendix N). The international character of the event was very important to create a flow of collaboration between Calabria and research centers in Europe. The activity of CRAI was growing rather fast and Sergio De Julio had started to promote various kinds of international actions to introduce his center in international projects. The database seminar, among other initiatives, definitely fit in with such ambitious goals.

At the same time in Rome we were trying to follow the most recent developments in the field of algorithms and complexity. The field was evolving and, as I had perceived during my trip in the US, the domains of computational geometry and parallel computing were starting to dominate the scene: two subjects that marked the late 1970s and the early 1980s. The IBM European Center for Scientific and Engineering Computing in Rome wanted to organize a seminar on parallel computing in 1982 and asked for the collaboration of our department and of CNR. In that period IBM had started to consider supercomputing as a major mission of the company. At the end of the 1970s ILLIAC IV and CDC STAR-100 had been practically the only powerful parallel computers (although constantly challenged by the technological advances achieved in the design of single processor machines) but in the 1980s the competition among computer companies for the construction of parallel computers was becoming stronger and various kinds of machines, based on different principles and on different interconnection topologies, were built as prototypes or as real commercial products (such as the Fujitsu vector supercomputers, the IBM 3090 with Vector Facility, the Connection Machine

---

[12] Lucio Bianco (born in 1941) would be nominated Director of the CNR Project on Transportation Systems in 1981 and from 1997 until 2003 he was President of CNR.

or the transputer-based massively parallel computers). We accepted with enthusiasm the proposal of IBM. The subject of the seminar was at a crossroad between theoretical issues and practical, technological advances and would have given us the possibility to bring to Rome some of the major figures of that new important field of research.

In February I took part in a workshop on Efficient Algorithms in Oberwolfach, organized by Kurt Mehlhorn[13], where the two above-mentioned domains of computational geometry and parallel computing were confirmed to be the most 'trendy' topics. Clearly classical subjects in algorithm design continued to be addressed by several authors: in particular the design of efficient data structures (talks of van Leeuwen, Flajolet, Ottmann, Munro, Paterson among others) and optimization and approximation (including Monien, Papadimitriou, Eli Shamir, van Emde Boas and myself); a few papers were also devoted to automata and formal languages (Kemp, Slisenko). But the two domains of computational geometry and parallel computing attracted a lot of interest. In the first field various papers were using the sweeping line method (invented two years before by Bentley and Ottmann) for solving geometric problems such as computing properties of sets of rectangles (Derick Wood) or the intersection of simple polygons (Herbert Edelsbrunner) or adjacency, connectivity and a wide range of other problems (Jürg Nievergelt). Besides some papers addressed the design of multidimensional data structures, suitable for geometrical applications (Hans-Peter Kriegel and Ed McCreight). Parallel computing was also addressed by several participants under different points of view. The main issue was clearly (and provocatively) stated by Les Valiant who asked the question whether general purpose parallel computing is at all possible. To this question Les Valiant gave a reassuring answer showing that it is possible to implement an 'idealistic' parallel computation model with N processors and shared memory on a 'realistic' architecture like an N node hypercube with a processor on any node with only $\log N$ time overhead. Positive answers to the question were also given by Wolfgang Paul and by Friedhelm Meyer auf der Heide who discussed the overhead deriving from the adoption of specific interconnection patterns as general purpose architectures. Other aspects of parallel computing were addressed by Wolfgang Paul (communication costs), by Erik Schmidt (synchronization) and by John Savage and Jon Bentley (VLSI design). Finally also Kurt Mehlhorn (Fig. 9.6) presented a talk somewhat related to parallel computing in which he discussed the cost to embed a graph in another graph (or, in other words, the

---

[13] Kurt Mehlhorn, born in 1949, is one of the leaders of European research in algorithms. He earned his PhD at Cornell in 1974 under the advisorship of Bob Constable and in 1976 he was already Chair of the Computer Science Department of Saarland University. Later, in 1990 he became Director of the Max Planck Institute for Computer Science and in this position he has made outstanding contributions in many areas of algorithmic research (data structures, computational geometry, parallel computing, combinatorial optimization) and in particular he has promoted the new domain of algorithm engineering.

Fig. 9.6 Kurt Mehlhorn

cost to implement an interconnection network on a different interconnection network).

In Europe in that period the relations among research groups in theoretical computer science of various universities were becoming more intense and more formal and this was without doubt one of the positive consequences of the existence of EATCS that was creating a lively environment for meetings and exchanges of visitors. At the beginning of the 1980s the Polytechnic University of Catalonia (Universitat Politècnica de Catalunya, UPC) had started to assume an important role in theoretical computer science. At that time the Facultad de Informatica was organized in several departments, among them the Department of Theoretical Computer Science, which included Marti Verges, Rafel Cases, Josep Estrada and Josep Díaz[14]. Besides, algorithms and complexity issues were also addressed in the Department of Operations Research in which were working Berenguer and Barceló[15]. In 1980 and 1981, through a series of visits in Paris and in Rome, Josep Díaz and his colleagues took several initiatives for establishing formal scientific collaboration ties with Italy and France. In particular Maurice Nivat and Josep Díaz signed a *Pro-*

---

[14] Born in 1950 Josep Díaz obtained his PhD in Valencia in 1981 (under the advisorship of Isidro Ramos) and has been one of the driving forces who played an essential part in the building and development of theoretical computer science in Spain. Through his great scientific work and his role as an energetic member and President of EATCS Josep has also given a very strong contribution to the growth of theoretical computer science in Europe. One of the most remarkable students of Díaz has been José Balcazar. Josep has strong Catalan feelings: in one of his letters to me he referred to the central Spanish state as "the Kingdom of Castille"!

[15] Later the Polytechnic University of Barcelona merged various departments (Programming Languages, Computer Vision and Theoretical Computer Science) into the Department of 'Llenguatges i Sistemes' which also included artificial intelligence and which for several years was the main research institution in informatics at UPC.

jet de cooperation Hispano-Française en Informatique Théorique in which the following topics were contemplated: theory of languages and of automata (involving Maurice Nivat and Dominique Perrin on the French side and Ernesto Garcia Camarero on the Spanish side), algorithmics (involving Philippe Flajolet and Josep Díaz), theory of programming (involving Maurice Nivat and Fernando Orejas), parallel and distributed systems (involving Claude Girault and Isidro Ramos), and programming and compiling (involving Michel Scholl and Fèlix Saltor). For any topic in the agreement a program of courses to be delivered in Spain and in France was foreseen, together with the possibility to enroll Spanish students in some French universities for graduate studies. A similar agreement was signed between UPC, our department in Rome University and the newborn Institute for Systems Analysis and Informatics of the Italian Research Council.

During the Spring of 1981, beside exchanging visits with our Spanish colleagues, we had a long series of other visitors in Rome: Philippe Flajolet, Bernhard Korte, Jan Karel Lenstra, Maurice Nivat, Alexander Rinnooy Kan. In view of the 1982 IBM seminar on parallel computing we also invited to Rome two of the leading figures of this domain, Hsing-Tsung Kung, whose work on systolic architectures was mentioned already and will be better explained in the following, and Franco Preparata, engaged at that time in problems related to the design of VLSI chips.

In 1981 ICALP was organized in Israel. Apart from the scientific interest of the conference this was a wonderful chance to visit again Israel, after ten years, and meet Israeli friends and colleagues. Anna had never been to Israel before and therefore she was eager to come with me; furthermore the conference was going to take place in Acre (Akko), the ancient Saint Jean d'Acre, which had played such an important role in the Middle Ages, at the time of the Crusades, and this added also a professional interest for her.

The conference was chaired by Shimon Even and Oded Kariv. Again a view of the major topics addressed in the conference may be interesting. Out of 43 papers presented at the conference 9 were devoted to formal languages and automata while 7 concerned various logics of programs (in particular dynamic logic was studied in two papers by Albert Meyer and Grazyna Mirkowska and by Joseph Halpern and Amir Pnueli) and 6 concerned algorithms and complexity issues (including a paper by Preparata and Vuillemin on VLSI networks for computing integer multiplication, and a paper by Mehlhorn and Rosenberg on graph embeddings). Other topics addressed at the conference were abstract data types and database theory. In particular to database theory was devoted Jeff Ullman's invited talk. In his talk he illustrated "the most fruitful research directions" that, in his view, included "... support for the universal relation concept ... and an exploration of the properties of acyclic database schemes". The concept of *universal relation* was indeed a popular, hot topic at that time. It had been introduced in 1978 by Beeri, Bernstein and Goodman in their paper *A Sophisticate's Introduction to Database Normalization Theory* and was based on the assumption that "... all relations in

Fig. 9.7 Jeff Ullman

a database are projections of a single relation". The idea of the universal relation was highly debated in the database community (some specialists even spoke of a "universal relation war"). Ullman (Fig. 9.7) was rather in favour of this point of view and in his talk he presented various positive aspects: "In order to make queries that involve data from more than one relation, it is convenient to imagine that the true world these relations represent is a single relation whose scheme consists of all the attributes mentioned in the scheme for any of the relations". Besides Ullman, other supporters of the universal relation considered that this model provided also a simplified form of user interface to the database. Also the issue of cyclicity of database schemes[16], that was particularly significant in the context of the universal relation approach, was discussed in Ullman's survey. Recently various authors had defined notions of acyclicity that could guarantee the consistency of a set of relations. In the seminal paper *Properties of Acyclic Database Schemes*, presented at STOC 1981, Beeri, Fagin, Maier, Mendelzon, Ullman and Yannakakis had pointed out some advantages deriving from acyclicity and more advantages were illustrated in another paper in preparation: *On the Desirability of Acyclic Database Schemes* by Beeri, Fagin, Maier and Yannakakis (a fundamental paper that now counts more than 800 citations, it appeared in 1983 on the Journal of ACM).

The stay in Acre was very pleasant. Walking through the very well preserved buildings from the Crusader's time like the Refectory of the Hospitaller's Knights was a real pleasure. Particularly interesting for us Italians were the traces of the presence of the Italian maritime republics Venice, Pisa and Genoa who in Acre had their own districts and marketplaces and who never stopped fighting each other even when they were in the Holy Land. In June, only one month before ICALP, there had been clashes at the Lebanon border, not far from Acre, between Palestinians and Israeli forces and the

---

[16] Given a database scheme we can define a hypergraph in which each hyperedge consists of the set of attributes of a relation. Essentially a database scheme is cyclic if the corresponding hypergraph is cyclic (for example the database scheme $\{(A, B, C), (C, D, E), (E, F, A)\}$ is cyclic).

tension could be perceived in the city; young women soldiers were patrolling the beach in short pants, armed with machine guns. The classic conference excursion was fantastic but hellish: the Dead Sea and Masada under a terrible sun. The only way to survive was to wear a wet fresh towel on the head.

The conference gave me also the occasion to spend some time with the Italian colleagues who were taking part in ICALP (Franco Preparata, Paolo Sipala and many others). In particular it was for me a good chance to make a better acquaintance with Alberto Apostolico (Fig. 9.8) who was traveling with his mother Rosa. I had met Alberto on various occasions (probably the first time had been at ICALP in Udine) and it had been always a pleasure to talk with someone who was embedded in such a rich and warm Southern Italy culture. On that occasion Alberto introduced me to his scientific interest, in particular to the new area that he was starting to explore: algorithms on strings. In fact just in those days he was working with Franco Preparata on the problem of computing subword repetitions in words, a study that after two years led to the publication of one of the most successful of Alberto and Franco's papers, *Optimal Off-line Detection of Repetitions in a String*, which now counts more than 250 citations[17].

Fig. 9.8 Alberto Apostolico at the Maratea Seminar in 1984

---

[17] Alberto Apostolico (1948–2015) was an outstanding computer scientist. He was a professor in various Italian universities and subsequently at Purdue and Georgia Tech. He was one of the leaders of the field of stringology, the field of analysis and design of algorithms for the most elementary type of data in computing, strings (or words): finite sequences of characters from a finite alphabet. The field was named 'stringology' by Zvi Galil who together with Alberto organized one of the first and most successful events concerning this domain: the *NATO Advanced Research Workshop on Combinatorial Algorithms on Words* held in Maratea, Italy, in 1984. Alberto's work covered most of the aspects of this field, from the design of efficient data structures for string representation (his most cited paper, with more than 400 citations, *The Myriad Virtues of Subword Trees*, appeared in 1983) to combinatorial pattern matching and to computational biology. In the words of Gadi Landau: "Alberto's contributions to the scientific world are enormous" (*www.cc.gatech.edu/sites/default/files/apostolico*).

At the beginning of September I went to Cetraro, in Calabria, for the Advanced Seminar on Theoretical Issues in Databases (TIDB) organized by CRAI. As it was said before, the seminar was chaired by François Bancilhon and myself with the collaboration of Mimmo Saccà and Nicolas Spyratos. The reaction of the database community to our invitations had been wonderful. With the exception of Jeff Ullman who had other commitments, all the scientists we had invited were there. The program was outstanding. The core topics in relational database theory were addressed by François Bancilhon, who spoke about database mappings, Mimmo Saccà, Catriel Beeri and Yehoshua Sagiv (from the Hebrew University of Jerusalem), who spoke about the implication problem for data dependencies, about the universal relation model, and about equivalence and optimization of relational expressions, Ron Fagin (from San José IBM Research Lab), who spoke about acyclic database schemes (cf. Fig. 9.9). Mihalis Yannakakis (from Bell Labs) and Christos Papadimitriou[18] (at that time at MIT) lectured about concurrency control in databases.

Fig. 9.9 Ron Fagin

Seymour Ginsburg (from USC) illustrated his attempt to extend to families of functional dependencies the algebraic approach that he had successfully applied to abstract families of languages. Carlo Zaniolo (from Bell Labs) spoke about incomplete database information and null values. Jan Paredaens (from Antwerp) spoke about horizontal decomposition of relations. John Mylopoulos (from Toronto) spoke about semantic models, Witold Lipski addressed the problem of extending the algebraic approach to database query. Finally Bob Kowalski's lectures concerned logic as a database language. Kowalski

---

[18] Christos Papadimitriou is an outstanding scientist who has made fundamental contributions in several diverse domains of theoretical computer science, ranging from optimization, approximation and on-line algorithms to game theory and from database theory to algorithmic aspects in the life sciences. Born in 1949 he has been a professor at Harvard, Athens, Stanford, San Diego and, currently, Berkeley. He is one of the most highly cited computer scientists. Among his students are Paris Kanellakis, who unfortunately died in an airplane accident, Constantinos Daskalakis, Elias Koutsoupias and many other excellent computer scientists. Some of his books such as *Elements of the Theory of Computation* (with H. R. Lewis), *Combinatorial Optimization* (with K. Steiglitz) and *Computational Complexity* are among the most popular textbooks in the history of the theory of computing.

Fig. 9.10 Bob Kowalski

(Fig. 9.10) was at Imperial College and his book *Logic for Problem Solving* had appeared two years before. Logic applications in databases were also addressed by Jean-Marie Nicolas; the application of logic programming to databases (which would become one of the major research directions both in databases and in artificial intelligence throughout the 1980s) was in its initial stage[19].

Beside the main lectures short talks were also given by Italian scientists such as Daniel Bovet, who spoke about concurrency control, Carlo Batini, who spoke about equivalence between database schemata and about the CNR project DATAID. A Polish database professional who had been hired at CRAI, Witold Staniszkis, lectured about database design for Codasyl systems. The seminar was attended by about 80 young researchers. Some of them (e.g. Maurizio Lenzerini and Paolo Atzeni, at that time still post-docs) later became leaders in the field. In the report that he wrote for the EATCS Bulletin, Christos Papadimitriou (Fig. 9.11) says: "There was an extensive mosaic of topics covered, reflecting the state of wild fermentation in which database theory appears to be. Relational theory, concurrency control, and logic in databases were three areas that were represented quite heavily in the list of speakers. Voices of the null value problem, schema design theory, the operational approach, conceptual modeling, and software engineering issues, were also heard". He also refers to disputes regarding different points of view on hot topics such as the universal relation: "As always there was exchange of results and views, amicable but lively disagreement and criticism, and the usual amount of misunderstanding and harassment over the universal relation assumption!"

Although we should realize that the success of the seminar was also due to the marvelous location ("unbeatable setting", in the words of Christos) and to the relaxing schedule that left time for swimming every day from 2 to 4 p.m. (we still remember Christos and Mihalis proving theorems and writing notes while sitting in the sea water), the scientific quality of the event was at the top level. Mimmo Saccà still has fond memories of the scientific work that was done at the seminar: "I still remember that during the seminar

---

[19] As we said before the most important database language inspired by logic, Datalog, had been introduced at about the same time.

Fig. 9.11 Christos Papadimitriou

Christos Papadimitriou gave a contribution to Fagin and Casanova's results providing a precise complexity analysis of some decision problem for inclusion and functional dependencies and for this reason was added as coauthor of the paper that was presented at the first PODS conference in May 1982". The impact of the seminar in the first steps of the field of database theory was remarkable. At the 2011 celebration of the 30th anniversary of PODS (the Symposium on Principles of Database Systems that was started in 1982) two events were remembered as progenitors of PODS and for contributing to the birth of the field of database theory: the XP series of Workshops on Database Theory (XP1 took place at SUNY Stony Brook in 1980 and XP2 took place at Pennsylvania State University in 1981) and the TIDB Advanced Seminar of Cetraro.

Also the social atmosphere was very pleasant. On that occasion, together with my wife, I made acquaintance with many foreign colleagues starting longstanding friendships, in particular with Christos Papadimitriou and with Mihalis Yannakakis (Fig. 9.12) and his wife.

Again Mimmo Saccà remembers funny episodes: "During the dances that followed a memorable dinner at the swimming pool Catriel Beeri and I started dancing a wild *tarantella* and Christos Papadimitriou was pushed to an unexpected immersion in the pool.". On the weekend the excursion brought all

Fig. 9.12 Mihalis Yannakakis

participants to Reggio Calabria's Archeological Museum where it was possible to see the two gigantic Greek statues, known as the Bronzi di Riace, that had been found a few years earlier in the sea off the Calabrian coast. One evening a magic musical encounter kept all participants fascinated: the Portuguese guitarist who was playing in the evening for the hotel customers was joined by Anna Chagas Bovet in a long series of moving Portuguese and Brazilian songs. As Christos says in conclusion of his report in the EATCS Bulletin: "All said, it was one of the most successful meetings I have been to, certainly the most enjoyable".

In any case the field of relational database theory was blooming in the early 1980s. In March 1982 the first *ACM Symposium on Principles of Database Systems* was held in Los Angeles. The conference was chaired by Jeff Ullman and Alfred Aho. The papers presented at the symposium (37 papers selected out of 104 submitted) addressed a large spectrum of topics: the universal relation model (still a debated issue: the title of Ullman's talk was *The U.R. Strikes Back*), cyclic and acyclic schemes, concurrency control, algebraic and logical aspects of databases, data dependencies, distributed databases and physical data structures. A few months later, in December 1982, in Europe another important event took place. A workshop was organized in Toulouse by Hervé Gallaire, Jack Minker and Jean Marie Nicolas, with the support of the Centre d'Etudes et de Recherches de l'Ecole Nationale Supérieure de l'Aeronautique et de l'Espace de Toulouse (C.E.R.T.). The title of the workshop was *Advances in Database Theory* and it gave rise to the publication of the book *Advances in Data Base Theory – Volume 2*[20]. The five major topics addressed in the book are: database schema design, integrity constraints, incomplete information, abstract data types for formal specification, theoretical aspects of query languages. In the introduction the editors claim that the field of databases is still in a state where many practical questions need to be answered (how does one choose a type of schema, how does one express and handle integrity constraints, how does one deal with real-world properties such as the incompleteness of information, etc.) and state what in their view were, at that time, the major theoretical instruments that could be used for addressing such practical issues. "It is not surprising that with the many problems that need to be solved, alternative formalisms need to be introduced. Mathematical logic is one such formalism. It is certainly the one most used as it is applicable to many database problems, not only that

---

[20] The book was actually the third in a series of three books and was preceded by two books devoted to previous workshops organized in Toulouse by the same group of people. The first one *Logic and Databases* (also known as *Advances in Data Bases – Volume 0*) contains the proceedings of the workshop *Logic and Data Bases* held in 1977 and is entirely devoted to logical aspects of databases (such as logic and integrity constraints, deductive question answering, knowledge representation, predicate calculus and query languages). The second one, *Advances in Data Bases – Volume 1*, contains the proceedings of the workshop *Formal Bases for Data Bases* held in 1979 and devoted to more classical aspects of database theory.

of developing a deductive capability for a database[21]. Other formalisms that have been used include algebra and the theory of hypergraphs."

In 1984, three years after the Cetraro Seminar, François Bancilhon and Nicolas Spyratos decided to organize another seminar on *Theoretical Issues in Databases*, in Benodet, on the coast of Bretagne: TIDB '84. Although various topics addressed in the seminar were similar to the topics addressed in Cetraro two new subjects appeared in the seminar: database machines, as we already said special purpose computer systems used as back end for quick access to large data repositories (a field in which especially Silvio Salza and Michel Scholl were at that time active), and geographic information systems.

Almost at the same time, together with some of the most active European researchers in databases, Serge Abiteboul, François Bancilhon, Jan Paredaens, Domenico Saccà, Nicolas Spyratos, we thought that the field deserved a permanent European forum and decided to organize the first European conference explicitly devoted to database theory: the *International Conference on Database Theory* (ICDT). It would be organized in Rome in 1986 and we chose the month of September not to overlap with the ACM PODS conference. Beside the above mentioned colleagues the program committee of ICDT '86 included Alessandro D'Atri, Marina Moscarini, John Mylopoulos, Jean-Marie Nicolas, Jurg Nievergelt, Christos Papadimitriou, Jeff Ullman, Moshe Vardi. The conference was rather successful and became a permanent event, one of the major events in the field[22]. We selected 24 papers covering classical database theory topics (data dependencies, acyclic database schemes, integration and decomposition of relations, etc.), deductive databases and other logical aspects in databases, concurrency control, physical data structures (hashing, list organizing strategies). One paper addressed aspects of semantic data models and object oriented databases, a subject that would become central in database systems and database theory in the 1990s. At the conference Laks V.S. Lakshmanan was awarded the best student paper prize named after Witold Lipski who had died the year before.

But let's go back to the Fall of 1981. After the Cetraro seminar I was soon involved in the commitment we had taken on with the IBM Rome Research Center: the scientific planning of the seminar on parallel computing. At the same time our visitors program continued with the aim to bring to students and young researchers in Rome some of the leading figures of algorithmic research in Europe. In October and November Burkhard Monien spent two months in our department. At that time Burkhard was switching

---

[21] The history of the interaction between logic and databases spans over 45 years. In his 2007 book *Logic and Databases* Chris Date, one of the pioneers of the relational database model, says: "Logic and databases are inextricably intertwined" (see also Phokion Kolaitis' Simons Institute lectures on YouTube).

[22] ICDT is still one of the major conferences in databases. Since 2009 it is held annually in conjunction with the conference *EDBT (Extending Database Technology)*. Another conference devoted to theoretical issues in databases was started in 1987, in Dresden (DDR) with the name *Mathematical Foundations of Database Systems* (MFDBS) but in 1992 this conference merged with ICDT.

from formal languages to complexity and was working on approximation algorithms for the vertex cover and other combinatorial optimization problems with his PhD student Ewald Speckenmeyer. A few years later Burkhard and Speckenmeyer would produce one of their most cited papers concerning algorithms for the satisfiability problem (340 citations). Monien had also started his longstanding collaboration with Hal Sudborough which would produce several important papers concerning graph algorithms. The visit in Rome of Burkhard and his wife Edeltraut was a pleasant event that was repeated an uncountable number of times, almost every year since then, marking a deep fondness between our families.

The last commitment that year was a trip to Paris in December. I was invited to INRIA by Flajolet. As usual the lively scientific atmosphere of the INRIA laboratories was a great source of new ideas. At that time Philippe was intensely working on combinatorial analysis of data structures with deep mathematical techniques. His paper on the analysis of possible labelings of paths in the plane with a method based on continued fractions appeared in 1980 and has been one of his most successful publications with now more than 400 citations. That was also, again, another nice occasion to see many old friends: Maurice Nivat, François Bancilhon and Jean-François Perrot (who had just moved to his new apartment on the Ile Saint-Louis). Andrzej Salwicki (Fig. 9.13) was also visiting Paris in those days, exactly at the moment in which General Wojciech Jaruzelski had seized power thanks to a military coup aimed at stopping the democratic wave raised by the Solidarność movement. I still remember the words of Andrzej, full of anxiety, saying that he was feeling seriously threatened by the situation in his country.

In the subsequent winter my engagement with CRAI brought me at least a couple of times per month to Calabria. Traveling to Calabria in that period was not easy. The airline company Itavia in fact was in a bad financial situation and flights were often canceled at the last moment if the number of

Fig. 9.13 Andrzej Salwicki and Grazyna Mirkowska in 1981

passengers was too small to cover the expenses (Itavia in fact went bankrupt a few years later). Besides during the winter the highway leading from the airport of Lametia to Cosenza was often interrupted by snow. Still the enthusiasm for the adventure of CRAI, for the objective to create an advanced computer science research center in Southern Italy, was so exciting that no difficulty could stop either me or the consultants who were coming from other Italian universities. The success of the seminar in Cetraro further encouraged CRAI's President Sergio De Julio to pursue his idea to develop forefront theoretical research as a scientific basis for innovation in computer applications. So the next problem was to hire a consistent number of computer specialists able to advance such ambitious goals but the problem, at that time frequently encountered by Italian software companies, was that in Italy the number of graduates in computer science was still limited, especially in Southern Italy, and it was difficult to find good people to hire. In particular it was difficult to find enough graduates with a suitable professional profile, the right mix of practical programming expertise, competence in formal methods of software design and intellectual curiosity for foundational aspects that make a good researcher in computer science. So De Julio decided to apply for European funds to run a homemade three years education program in computer science for selected graduates (no matter in what discipline) and asked me to lead this project.

I started to assemble a board of about ten colleagues from the major Italian computer science groups (among others Ugo Montanari, Giorgio Levi, Carlo Ghezzi). All of them greedily accepted and we started to meet in Calabria at the beginning of 1982 in order to design the curriculum. In a sense organizing a three years computer science course was for me (and for the colleagues that I involved in the project) like organizing a new unprecedented university curriculum in which, at last, we had the chance to give the due room to the theoretical foundations of computer science. At a time in which the Faculties of Engineering did not yet have (as they have now) a degree in computer science (and in any case no degree in computer science was offered in Italy south of Salerno and Bari) the idea to recruit 30 bright graduates in various disciplines (chemists, biologists, mathematicians, etc.) and to turn them into computer science specialists able to take part in advanced research projects was really exciting.

The collaboration with the project of Mario Bolognani, at that time director of Systems and Management (S&M), one of the few Italian companies that were familiar with formal methods in software design, allowed us to match the scientific program of the school with the needs of innovation in the design of computer systems and computer applications. The idea on which the course was based is still unique in its philosophy: the first year of the course was strongly inspired by software engineering principles. We decided to follow the phases of what, at that time, was the typical software product life cycle, from user requirements to software specification, coding, testing and debugging. For any phase we wanted to introduce the students to the most advanced formal methods and to the use of design support tools. At the same time, in parallel, we had to teach various basic topics in computer science education: theoretical foundations of computer science (mainly computability and complexity), formal languages and compiler construction, computer system organization and operating systems (from mainframes to local area networks of workstations). The second year of the program was completely devoted to the development of projects of educational nature, in which all software engineering concepts that had been taught during the first year would be put into practice. During the third year, beside providing some specialized monographic courses to complement their education, the students were included in CRAI teams for the development of real projects (ranging from simple applications to more demanding research projects). The course started in May 1982. The formula that we had chosen for the course was indeed very successful and three years later it was replicated for hiring thirty more specialists at CRAI. Most important, the formula was adopted again, although with a reduced time scale of one year instead of three years, by a fairly large Italian software company, ENIDATA, for which CRAI organized three courses in the following years, always with the target of converting bright young graduates in other disciplines into computer specialists.

Chapter 10
Europe Strikes Back

The IBM *Seminar on Parallel Computing* took place in Rome at the IBM European Center for Scientific and Engineering Computing in March 1982. The subject, as stressed already, was very up to date and the seminar attracted many Italian researchers and professionals. On the IBM side the organizers were Giorgio Sommi, the Director of the Center, and Piero Sguazzero, an expert on scientific computing on IBM vector machines. Traditionally IBM was the main provider of computing power especially for the communities interested in scientific computing (physicists, mathematicians, biologists) and they wanted to maintain their role in front of other providers of supercomputing facilities that were starting to compete (such as for example Silicon Graphics, founded in 1982), making clever use of technological advances such as pipelining the execution of instructions or extending the computer power with specialized processors (a mathematical processor or a graphic processor).

As already pointed out the subject of parallel computing has been present in the theoretical computer science literature since the 1960s and early 1970s, with the first projects for the design of multiprocessor machines. Indeed one of the most cited papers in the field (more than 2400 citations) is Michael Flynn's paper *Some Computer Organizations and Their Effectiveness* (appeared in 1972 in IEEE Transactions on Computers), in which the basic classification of parallel computers in terms of SIMD machines (single instruction, multiple data stream) and MIMD[1] machines (multiple instruction, multiple data stream) is presented. While the SIMD machines somewhat corresponded to practically feasible architectures such as the array or vector processors (in which the same operation is executed in parallel on all the elements

---

[1] Speaking about MIMD machines Italians have a motivation for being proud. In fact the first idea for a MIMD machine came to the mind of Luigi Federico Menabrea (general of the Piedmont kingdom and scientist) who in 1842 published the paper: *Notions sur la machine analytique de Charles Babbage* with the 'Bibliothèque Universelle de Genève', in which he envisioned the possibility to use several machines in parallel to perform complex tasks.

© Springer International Publishing AG, part of Springer Nature 2018
G. Ausiello, *The Making of a New Science*,
https://doi.org/10.1007/978-3-319-62680-2_10

of an array) the case with MIMD machines was less clear and gave rise to a variety of theoretical parallel computation models, depending on whether all processors were accessing the same common storage (possibly with exclusive read or write rights) or each processor had its own private storage and the interprocessor data communication was realized by messages flowing through the interconnection network. At the beginning of the 1980s a connection between different models of MIMD machines was established thanks to the result obtained by Valiant and already mentioned above. In the paper *Universal Schemes for Parallel Communication* presented at STOC 1981 (another landmark paper with more than 700 citations) Valiant had described and analyzed random routing (a method for reducing contention in networks) showing that it is possible to implement an 'idealistic' parallel computation model with shared memory (such as a P-RAM) on a 'realistic' architecture like a hypercube with a processor on any node with a small overhead.

Also various models allowing some kind of *massively parallel computing* were introduced in the same period, such as for example the Connection Machine, very popular in the AI community[2], which started as a SIMD machine and later evolved into a MIMD machine, and the machines based on INMOS transputers.

Other parallel computation models, inspired by technological developments, were introduced in the same years also for SIMD-like machines and had become quite popular. In particular, the possibility to implement in VLSI (very-large-scale integration) very large arrays of specialized processors on a chip opened the road to the so-called *systolic arrays*. A systolic array[3] is a network of tightly coupled *data processing units* in which each unit receives data from upstream units, processes them, stores the result and passes data to downstream units. The flow of data resembles the flow of blood when it is pumped by the heart; the name came from this analogy.

All these aspects were presented at the IBM Seminar. The stars of the seminar were undoubtedly H.T. Kung (Fig. 10.1) and Franco Preparata. Both spoke about VLSI parallel computation models. Kung, of course, gave a great overview of systolic arrays and their applications along the lines of his landmark paper *Why Systolic Architectures?*, which appeared in the same year and now counts more than 2500 citations. Preparata spoke about optimal implementations of various arithmetic algorithms in VLSI and also about aspects related to VLSI design such as the so-called *channel routing* problem, that is the problem to find an optimum layout for the interconnections be-

---

[2] The first design of the connection machine was provided in 1981 by Danny Hillis at the MIT Artificial Intelligence Lab and the machine was produced, in 1983, by the company Thinking Machine. It consisted of several thousand (up to 65,000) one-bit processors connected in a hypercube topology. The machine could be programmed in a version of LISP.

[3] One of the first descriptions of systolic arrays was given by H.T. Kung and C.E. Leiserson in the chapter *Algorithms for VLSI Processor Arrays* that appears in Mead and Conway's 1979 book *Introduction to VLSI Systems*.

Fig. 10.1 H.T. Kung

tween processors and memories on a chip. Another notable speaker was Ken Kennedy, from Rice University, who spoke about PFC, the Parallel Fortran Compiler. Classical SIMD models such as vector machines and their applications were presented by various IBM specialists. Finally Giacomo Cioffi and I gave a broad and historical introduction to parallel computation models and in particular to MIMD models.

Since then, task parallelization on suitable architectures has been widely applied in practice in order to better deal with time-consuming processing of large data sets, with different approaches applied under different conditions. A context where the impact of parallel computing is now especially relevant is the one given by highly-responsive processing, such as for example real-time graphics processing or video decoding: here, specialized *Graphic Processing Units* (GPUs) relying on SIMD parallelization are introduced and integrated in different possible ways with the Central Processing Unit.

The computing speed of GPUs can be exploited also in different frameworks where SIMD parallelization turns out to be profitable, such as in many scientific computing tasks. In particular, the wide and successful application of large and complex neural networks to machine learning, known as *deep learning*, is based on the use of GPUs, both single and in clusters.

MIMD-type parallelization approaches are even more widely applied, for example through multi-core processors, where two or more independent processing units may execute different software in parallel, or through integrated computer clusters, that is collections of computing nodes connected through fast switches or LANs and orchestrated in a centralized way. The use of very large clusters in an integrated framework for data storage and processing, such as for example the Hadoop framework, makes it possible to process and extract information from the very large (terabyte, petabyte) datasets currently available.

After the summer, a wonderful summer that Anna and I spent on the Turk-
ish southern coast, at Bodrum, the ancient Alikarnassos, a series of events
were waiting the European algorithms community.

First of all, at the end of August 1982 came the XI International Symposium on Mathematical Programming that was being held in Bonn, organized
by the Institut für Ökonometrie und Operations Research of the University of Bonn, Korte's institute. The symposium was strongly marked by the
overwhelming personality of Bernhard Korte and gathered the world's top
scientists working in the fields of operations research and combinatorics. An
impressive record of the relevance of the symposium is given by the book
Mathematical Programming – The State of the Art (edited by A. Bachem,
M. Grötschel and B. Korte) that contains 21 survey papers presented in Bonn
and provides a very comprehensive view of the state of the art of the field in
1982.

Then came the workshop on *Analysis and Design of Algorithms in Combinatorial Optimization* that Mario Lucertini and I had organized in Udine
with the collaboration of CISM. The workshop was a follow up of the series
of events on the same topic, the 1976 *Amsterdam Conference on Interfaces
Between Computer Science and Operations Research* and the two schools
held in Udine, in 1979, and in Barcelona, in 1980, and was again aimed at
strengthening the cooperation between experts in discrete mathematics, in
theoretical computer science and in operations research. In fact at the workshop a series of papers were presented on hot topics in all such domains, most
of which were published in the proceedings that appeared in 1985, as the 25th
volume of *Annals of Discrete Mathematics*. Particularly interesting to understand the breadth of the subjects covered in the workshop are the papers by
Ion Filotti on graph embeddings into surfaces of higher genus, by Philippe
Flajolet concerning the average case performance of the trie data structure,
by Dorit Hochbaum concerning the k-center and dominating set problems,
by Bruno Simeone on quadratic Boolean equations, and by Franco Preparata
on channel routing for multiterminal nets. Shimon Even (Fig. 10.2) also attended the workshop. On this occasion he presented a paper that became a
landmark in the history of approximate solution of optimization problems: *A
Local-Ratio Theorem for Approximating the Weighted Vertex Cover Problem*
written in collaboration with Reuven Bar-Yehuda (more than 400 citations,
one of the most cited papers of Shimon Even).

In order to understand the lively activity in optimization research in Europe in that period it is interesting to remember another important event
that took place in Gargnano, on Lake Garda, soon after the Udine workshop, the *International Institute on Stochastics and Optimization* organized
by Francesco Archetti (who also edited the volume of proceedings published
by North Holland) and Francesco Maffioli. The event marked the growing
importance of probabilistic and statistical tools in the design and analysis of
algorithms for both discrete and continuous optimization problems. As the
announcement of the seminar says "there are three levels of involvement of

Fig. 10.2 Shimon Even

stochastics in the context of optimization: (1) the modeling of the problem; (2) the design of the algorithm; (3) the analysis of the performance of the algorithm". All such aspects were addressed at the seminar by the contributed papers and by the invited speakers (an exceptional list of outstanding scientists including Václav Chvatal, Richard Karp, Jan Karel Lenstra, Christos Papadimitriou, Michael Rabin, Alexander Rinnooy Kan).

In the summer of 1982 Alberto Marchetti had left Rome to spend one year in Berkeley with Richard Karp, and I started to make plans to visit him in October. Also Mimmo Saccà had left Cosenza in January to spend one year at IBM San José and I might visit him too. Then I could make a stop in Urbana, where I was invited by Franco Preparata to give a talk, and finally I could attend FOCS which was going to be held in Chicago from November 3 to November 5.

In Berkeley Alberto had started doing research under the supervision of Richard Karp and in collaboration with one of Karp's students, Mike Luby, and with Andrew Goldberg (at that time a master's student)[4]. Eventually the research work, addressing two fundamental problems in combinatorial optimization, the multidimensional bin packing and the knapsack problem with zero-one variables, led to two STOC papers in 1984. The atmosphere in Berkeley was very pleasant and friendly as usual. After meeting again Manuel Blum, Richard Karp and Gene Lawler it was a big chance to meet with the young generation of stellar students: Silvio Micali, whose work in the mean-

---

[4] After the master's in Berkeley, Andrew Goldberg obtained his Ph.D. degree in Computer Science from MIT in 1987 under the advisorship of Charles Leiserson. In his outstanding research career, among other topics, Goldberg has addressed the design, analysis, and computational evaluation of algorithms and data structures, algorithm engineering, computational game theory, and electronic commerce. His algorithms are widely used in industry and academia. After working at Microsoft Research, currently Andrew Goldberg is senior principal scientist at Amazon.

time had achieved extremely good results giving rise to three papers accepted at 1982 FOCS; Vijay Vazirani[5], still working on his PhD thesis concerning maximum matching; his brother Umesh[6], another exceptional PhD student of Manuel Blum; and Michael Luby[7], a student of Richard Karp. During my stay in Berkeley I also visited IBM San José where Mimmo Saccà was carrying out research in relational database theory since January under the supervision of Ron Fagin and, part of the time, in distributed databases in collaboration with Gio Wiederhold.

Then I left for Urbana where I was again a guest of Franco Preparata and Rosamaria. Franco's research work was growing very successfully in the two fields of VLSI and computational geometry and several young brilliant researchers were attracted to join the faculty or to visit Urbana, such as Herbert Edelsbrunner, who had just obtained his PhD in Graz in computational geometry, under the supervision of Hermann Maurer, and Witold Lipski, with whom Franco had a paper concerning channel routing at the next FOCS. In 1982 another young Italian PhD student had joined the Department of Electrical Engineering, Gianfranco Bilardi, now a professor in Padua. It was also a nice surprise for me to meet with Alberto Apostolico and his wife Titti Guerra who were visiting the department in the same week. In 1979 I had invited Titti to move to our department in Rome which in fact she did one year later but, for now, she had decided to join Alberto in the US. Together with Franco, Alberto and Titti, a few days later we left by car, destination Chicago, to attend FOCS. The trip across Illinois was very pleasant; the whole state was painted with a superb foliage. The trip was also a good occasion to discuss with Franco and our companions about the decision to work and live in the US, a decision that Franco had made several years before and that Alberto and Titti were about to make, although later they would both become professors at the University of Padua.

FOCS 1982 is worth remembering as one of the most important conferences in the early 1980s. In particular cryptography had a major role at the conference and various papers introduced new concepts and new directions in that field. As I mentioned before, Silvio Micali had a successful presence at FOCS with three papers, one of them regarding matching in general graphs, in collaboration with Galil and Gabow, and two regarding cryptography. The most relevant, definitely a landmark in this research field, was the paper *How to Generate Cryptographically Strong Sequences of Pseudo Random*

---

[5] Vijay Vazirani later did excellent work on approximation algorithms. He would become first a professor at the Indian Institute of Technology, Delhi, later at Cornell University and then at Georgia Tech from where he recently moved to UC Irvine. Naveen Garg and Samir Khuller are two among his outstanding students.

[6] Umesh Vazirani devoted his research work to randomized computing and quantum computing and became a professor at UC Berkeley. Again he would collect a long list of exceptional students, among them Andris Ambainis, Sanjeev Arora and Madhu Sudan.

[7] After obtaining his PhD Mike Luby addressed himself to efficient coding techniques for multimedia delivery and streaming technologies, founded the company Digital Fountain and is now Vice President of Qualcomm.

Bits, co-authored with Manuel Blum, whose journal version has now more than 1500 citations. In contrast with Kolmogorov's notion of random strings, Blum and Micali for the first time introduced a totally new approach putting randomness in connection with computational resources and showing how to construct a polynomial-time algorithm that generates a string of pseudorandom 0-1 digits in such a way that no probabilistic polynomial-time algorithm can guess whether the next bit in the sequence will be 0 or 1 with higher probability than tossing a coin. Besides Micali's papers, two other important cryptography papers were presented by Andy Yao (Fig. 10.3): *Protocols for Secure Computations* (which has now more than 3000 citations) and *Theory and Applications of Trapdoor Functions* (more than 1700 citations). The latter paper addressed the subject of those functions that are easy to compute but hard to invert, unless the *key* for inversion (the trapdoor) is known. Trapdoor functions had been introduced in the original 1976 Diffie and Hellman paper *New Directions in Cryptography* but thanks to Yao's contribution they started to play a major role in cryptography in the 1980s.

Fig. 10.3 Andy Yao

But the relevance of the conference was not limited to cryptography. Also Christos Papadimitriou presented a fundamental paper in which he studied the complexity of finding instances with unique optimum solutions in optimization problems (in particular, how hard is it to establish that an instance of the traveling salesman problem has a unique optimum solution?). Other important domains addressed at the conference were linear programming (Nimrod Megiddo presented a linear-time algorithm for linear programming in R^3 with several consequences for geometric problems in R^3, a paper that now has more than 800 citations), VLSI (beside Preparata and Lipski's paper on channel routing particularly interesting was a paper by Tom Leighton and Charles Leiserson concerning fault problems in systolic arrays) and robotics (John Hopcroft had just started working in robotics and his paper at FOCS, written in collaboration with Deborah Joseph and Sue Whitesides, was one of the first contributions he gave on motion planning).

I still have a fond memory of my trip to Chicago. The stay in Chicago was indeed very pleasant and the city impressed me a lot with its modern style of architecture and the sculptures that embellished its squares. With Christos and his first wife Xanthippe we had a pleasant stroll in downtown Chicago and a fantastic visit at the Museum of Contemporary Art.

Back in Rome I was soon overwhelmed by the preparation of various events that had been planned for 1983: another workshop devoted to design and analysis of algorithms that we were organizing again in Udine and a conference that we had planned to organize in L'Aquila. As we have already seen, at the beginning of the 1980s the Italian theory of computing community was extremely active, and, in a sense, the fact that some well established conferences started to be organized in Italy was a sign that the community was being perceived as scientifically strong. In 1982 the major European conference devoted to semantics and the theory of computing, the *International Symposium on Programming*, that, as we have seen, had been initiated by Bernard Robinet and whose all previous editions had been held bi-annually in Paris, had been organized in Turin by Mariangiola Dezani. The Chairman of the Program Committee was Ugo Montanari. The topics addressed in the conference are representative of the studies carried on in this domain in that period: program verification, program transformation, concurrent programming, logic programming, semantics of nondeterministic programs, etc. Another series of conferences, that we already mentioned, initiated in France by André Arnold and Max Dauchet in 1976, had started in 1981 to circulate in other countries: the *Colloquium on Trees in Algebra and Programming*. Until then these conferences had been always organized in France, in particular in Lille, with the title *Colloque de Lille sur les Arbres en Algèbre et en Programmation* and with the acronym CAAP[8]. In 1981 the sixth CAAP conference was organized in Italy by Egidio Astesiano, the founder of a strong group in semantics at the University of Genoa, and by his colleague Gerardo Costa. Chairman of the Program Committee was Corrado Böhm. As the foreword of the proceedings says, "though the title looks restrictive, trees and related algebraic structures enter in almost every conceptual structure of Computer Science, so that *Trees in Algebra and Programming* includes a wide range of topics. A prominent feature of the presented papers is their mathematical character and so contributions related to algebra, mathematical logic and arithmetical complexity are also included". In fact the papers presented at the conference covered several topics both in the field of semantics and in the field of algorithms and complexity, extensively related with applications of trees and graphs in computing. The invited lectures concerned algebraic specification of data (Hartmut Ehrig), concurrency (Robin Milner), data and file

---

[8] Since 1985 CAAP was organized as a standalone conference in even years and, in odd years, as part of a larger event, called TAPSOFT, the *Joint Conference on Theory and Practice of Software Development*, in which the theoretical part represented by CAAP was supplemented with a *Colloquium on Software Engineering*. The first TAPSOFT was organized in Berlin in 1985 by Hartmut Ehrig, Christiane Floyd, Maurice Nivat and Jim Thatcher. Later, in 1998, as we have already said, the TAPSOFT conferences were absorbed (together with ESOP and other conferences) into ETAPS, the *European Joint Conferences on Theory and Practice of Software*.

structures (Jürg Nievergelt), and infinitary relations (Maurice Nivat). We had a paper on approximation of combinatorial optimization problems accepted in Genoa, and Marco Protasi, who at that time was an associate professor in L'Aquila, beside presenting our paper at the conference, presented a bid to hold CAAP in L'Aquila in 1983.

The 8th CAAP was held in L'Aquila in March 1983. Again the scope of the conference was extended to make room for some of the most important research topics in theoretical computer science at that time. As the foreword of the proceedings says, topics included "formal aspects and properties of trees and, more generally, of combinatorial and algebraic structures in all fields of computer science: theory of algorithms and computational complexity, formal languages and automata, theory of sequential and parallel programs, theory of data structures and data bases, algebraic specification of software".

The list of topics addressed in the invited lectures confirms this broad view: acyclic database schemes (Ronald Fagin), matching in graphs (Zvi Galil), graph grammars (Dirk Janssens and Grzegorz Rozenberg), heterogeneous algebras in data specification (Manfred Broy and Martin Wirsing), applicative systems (Mario Coppo, Mariangiola Dezani and Giuseppe Longo). The conference was rather successful and the list of contributors looks impressive (among them Christos Papadimitriou, Burkhard Monien, Philippe Flajolet, Don Sannella, Rod Burstall, Jean-Pierre Jouannaud, Sorin Istrail, Andrzej Lingas, etc.). On the last day we organized a fantastic tour of the most famous medieval castles and churches that surround the city, led by a professor of medieval history who was able to get all of us interested in the rich and fascinating history of the foundation of L'Aquila in the thirteenth century, under the rule of Friedrich II, when the crusaders used to travel through this territory to reach Brindisi and sail from there toward the Holy Land.

Somewhat in the same period, in the years 1982 and 1983, in Italy, a deep reorganization was taking place in the universities. All universities were being restructured by replacing the old organization, based on institutes, with a new system based on departments, larger aggregations of faculty members, corresponding to well identified research domains. Also the career system was being changed, with the introduction of three levels of professors, and moreover the new educational degree, the doctorate, corresponding to the PhD, was being introduced. Antonio Ruberti, at that time still Rector of *La Sapienza* University, put a lot of energy into making the reorganization as effective as possible. In particular in our field, in place of the old *Istituto di Automatica* a new department, called *Dipartimento di Informatica e Sistemistica (Department of Computer and System Sciences)*, was created. This was the first department of *La Sapienza* having computer science explicitly mentioned in its name[9].

---

[9] The new organization offered a great occasion to bring together all people working in computer science at *La Sapienza*, breaking the traditional barrier separating computer scientists working in the Faculty of Engineering and those belonging to the Faculty of Sciences. We made a serious attempt to invite Corrado Böhm, Daniel Bovet and other

The need to define a long term research program for the department was a great occasion of excitement for all of us. In a series of internal seminars we started to discuss and compare the broad spectrum of research directions that fit into the scope of the department, ranging from control theory to bioengineering, from algorithms to computer architectures, from artificial intelligence to databases, from discrete and continuous optimization to mathematical modeling in industry and utility systems.

The seminars provided a meaningful picture of the state of research in theoretical computer science in our department. In the area of semantics Alberto Pettorossi was carrying on his study of communication among processes in functional and applicative parallel programming, research work that he was carrying out in collaboration with the Polish logician Andrzej Skowron. In the field of databases two directions were dominating: algorithms and tools for the design of Entity-Relationship conceptual schemes and theoretical aspects of database schemes. The research activity in the first direction (which involved mainly Carlo Batini, Maurizio Lenzerini and Roberto Tamassia), although motivated by practical objectives related to the already mentioned DATAID project, triggered, in the subsequent years, a great number of superb theoretical results concerning graphical user interfaces for databases and graph drawing algorithms. The second direction (which involved Paolo Atzeni, Alessandro D'Atri, Marina Moscarini and myself) was more strictly related to the hot current topics in database theory (essentially combinatorial problems in databases: recognition and design of acyclic database schemes, equivalence, inclusion and integration among database schemes, minimal conceptual connections in acyclic dabases, etc.). The seminars were also a great chance to introduce to our colleagues Gigina Carlucci Aiello. Gigina had been recently recruited to our department, and she presented for the first time to the department her view of artificial intelligence research with a particular emphasis on knowledge representation. The subject of logic and functional programming was addressed by Alfonso Miola, who had also recently joined the department. Of course the subjects of design and analysis of algorithms for combinatorial optimization problems and of approximation algorithms still had a great role in our research work. In this field Mario Lucertini spoke about discrete optimization problems (graph partitioning, in particular) and I illustrated the research that I was carrying on with Alberto Marchetti Spaccamela (just back from Berkeley), Marco Protasi and a new very bright postdoc, Maurizio Talamo: the study of the performance of greedy algorithms for various optimization problems, both on general graphs and on random graphs, in order to understand when a greedy approach was suitable to reach approximate solutions with performance guarantee and why in other cases the approach failed.

---

colleagues still belonging to the *Department of Mathematics* to join us in the new organization but we were unable to overcome the opposition of the Dean of the Faculty of Sciences and we failed.

As we have seen, in those years the interactions between theoretical computer science and operations research were getting stronger. Mario Lucertini and I even wrote a paper, meant to be the 'charter' of the newborn CNR institute IASI, in which all research directions were in a sense organized around a core consisting of operations research, analysis and design of algorithms, and computational complexity. The organization of yet another event in Udine was therefore just another step in that direction. In July 1983 a school on *Algorithm Design for Computer System Design* was held in Udine under the sponsorship of CISM (the *International Centre for Mechanical Sciences*) and of the *Department of Computer and System Sciences* of Rome University. The school was chaired by Mario Lucertini, Christos Papadimitriou, Paolo Serafini (professor of operations research in Udine) and myself and the aim was to stress the need for formal and quantitative approaches in the design of complex computer (hardware and software) systems in order to address "such problems as optimal memory management, optimal design of computer networks and multiprocessors systems, optimal layout of VLSI systems, efficient exploitation of parallel computing systems, optimal management of database schemes, concurrency control, ...". Major lectures were given by Franco Preparata and Charlie Wong (on VLSI layout problems and on the design of a VLSI sorter, respectively), by Christos Papadimitriou (on concurrency control in databases) and by Ed Coffman (the paper *Approximation Algorithms for Bin-Packing – An Updated Survey* by Coffman, Garey and Johnson, included in the proceedings of the school, was for a long time the most cited paper on the subject, with more than 470 citations).

The presence of Christos at the school in Udine was not only important for the success of the event but especially for the flow of information concerning new ideas and new research directions that he was able to convey to the participants, in particular regarding the online management of dynamic information structures subject to sequences of modification operations and queries[10]. Following the results that had been achieved during the 1970s regarding dictionaries (sets of items supporting search queries and insertion and deletion operations), in the early 1980s the study of algorithms over more complex dynamic structures (e.g., sets, trees and graphs) had emerged as a new important research trend. One of the first contributions addressing this type of issue had been presented in 1975 by Robert Tarjan who had introduced efficient structures for manipulating a family of disjoint sets subject to *union* operations and *find* queries (*Efficiency of a Good But Not Linear Set Union Algorithm*, a paper that now has more than 1300 citations) A few

---

[10] A series of interesting open problems concerning online algorithms were presented and discussed in informal talks by Christos at the school. I carefully took note and transferred problems and ideas to one of my brightest students at that time in Rome, Pino Italiano. It is interesting to note that Pino devoted his master's thesis to the transitive closure of dynamic graphs (his first publication at the 24th Annual Allerton Conference on Communication, Control and Computing) and, since then, after obtaining a PhD at Columbia under the advisorship of Zvi Galil, he has been one of the major scientists who have contributed to the field of dynamic and online algorithms.

years later another paper by Robert Tarjan in collaboration with his student
Daniel Sleator was presented at STOC 1981 (*A Data Structure for Dynamic
Trees*, over 1000 citations). In this paper it was shown how to manage a forest
of vertex-disjoint trees subject to the operations of combination of two trees
in one tree and of split of one tree into two trees and for the first time the
notion of amortized cost (over a sequence of operations) was formally introduced. In the following decade a great number of papers started to address
problems on dynamic graphs, that is graphs (or weighted graphs) in which
edges could be inserted or removed (or edge weights could be increased or
decreased). Dynamic graph algorithms have a wide range of applications (especially in network management) and have become one of the major research
directions in the theory of computing.

In 1983 EATCS continued to be very active and to upgrade the service that
was offered to the international theory community. Grzegorz Rozenberg, who
was the editor of the Bulletin since October 1981, further enriched its content by adding new sections such as 'Books reviews', 'Reports on computer
science organizations' and 'Abstracts of Ph.D. theses'. Another important
initiative that was started in that period was the establishment of the book
series *EATCS Monographs in Theoretical Computer Science* in cooperation
with Springer, one of the most successful adventures in the life of the association. The editors of the series were Wilfried Brauer, Grzegorz Rozenberg
and the EATCS President Arto Salomaa, supported by an Advisory Board
initially consisting of Shimon Even, Maurice Nivat, Christos Papadimitriou,
Arny Rosenberg, Dana Scott and myself. The first volumes[11] rapidly became
fundamental textbooks in the field: Kurt Mehlhorn's *Data Structures and Algorithms* (in 3 volumes), Reisig's *Petri Nets*, Kuich and Salomaa's *Semirings,
Automata, Languages*, Ehrig and Mahr's *Fundamentals of Algebraic Specification*, etc.

Naturally the EATCS flagship conference was also becoming stronger. In
1983 ICALP was held in Barcelona. It was the 10th ICALP and also for this
reason EATCS gave to this event a particular emphasis but the event was
anyway quite memorable.

The conference was chaired by Josep Díaz (Fig. 10.4) and the scientific
program was extremely rich: 60 contributed papers (in 1981 there were 44
contributions and in 1982, 50) plus two invited lectures given by Michael Rabin and Jarik Nešetřil. One third of the papers were devoted to semantics and
logic, many of them addressing the use of temporal logic in the behavioural
analysis of concurrent programs (Apt, Arnold, de Bakker, Manna and Pnueli,
among others). Particularly interesting was the paper of Lehmann and She-
lah in which temporal logic was extended to take into account probabilities
of uncertain events. Various other aspects of the semantics of concurrent

---

[11] Ten books were published in the first four years. In the 1990s the series changed its
name and was transformed into two series: *Monographs in Theoretical Computer Science.
An EATCS Series* (currently includes 61 volumes) and *Texts in Theoretical Computer
Science. An EATCS Series* (currently 46 volumes).

Fig. 10.4 Josep Díaz

programs, including notions of equivalence of processes (De Nicola and Hennessy), a comparison between two models of concurrency, Milner's CCS and Hoare's CSP (Brookes), and the semantics of the rendezvous mechanism in the programming language Ada (de Bakker) were also considered.

About 25% of the papers were in the field of algorithms and complexity. Various papers concerned the analysis of data structures (2-3 trees, binary tournaments, etc.). A particular aspect of data structures that was arousing interest in those years was the concept of implicit data structures (also known as space-efficient data structures, that is, data structures in which the position of data elements carries meaning and very little space overhead is needed for pointers or other information constructs), introduced by Ian Munro in 1980. A paper on this subject was presented by Munro himself. An algorithm paper that deserves mention was presented by von zur Gathen and Kaltofen regarding polynomial-time algorithms for the factorization of multivariate polynomials over finite fields. Finally complexity papers were presented by Hartmanis, Vitányi, Schöning and Book. Most of the remaining papers were in the field of formal languages, automata, tree automata. Only one paper concerned database theory.

Beyond the scientific quality, the fame of ICALP 1983 is due to an incredible social program organized by our Catalan friends[12]. Thanks to the favorable political moment (after the end of General Franco's dictatorship the government of the city of Barcelona was eager to gain visibility at international level) and to the good relationships of Rafel Cases, one of Josep's

---

[12] After the Barcelona ICALP, another remarkable ICALP was held two years later in Nafplion (Greece), organized by Christos Papadimitriou and Stathis Zachos. The conference was chaired by Wilfred Brauer and invited talks were given by Lovász and Pnueli. A memorable event was the lecture given by Donald Knuth inside the Theater of Epidauros, concerning his work on the typesetting system TeX; the lecture was introduced by the pro tempore Greek Minister for Culture, the famous actress and politician Melina Mercouri.

colleagues, with Catalunya's political circles, the support received from the city of Barcelona was outstanding. The City Hall paid for an extra gala dinner for ICALP and let us use for free the *Drassanes Reials*, the late-medieval naval shipyard dating from the sixteenth century. At the center of the museum stood the gigantic flagship of Don Juan de Austria, a full size replica of the galley with which Don Juan led the Christian fleet to victory over the Turks at Lepanto. All tables were arranged on the bay around the ship. Very impressive. Another extraordinary dinner was held in the garden of the Royal Palace. Finally, a jewel that I am sure all participants have carefully preserved, the poster of the conference was designed by Juan Miró, one of his last paintings as he died at the end of that year at the age of 90.

During 1983, throughout the world, computer science was rocked by the thrust of technological advances. The year had opened January 3rd with the famous cover page of Time magazine on which, instead of the usual *Person of the Year*, an image of a desktop computer was presented under the headline: 'Machine of the Year – The Computer Moves In'. In fact, on one extreme end of the computing spectrum, after several years of experimentation on a wide variety of models (including, in the 1960s, the prehistoric Olivetti Programma 101 and, in the 1970s, the Xerox Alto, the Commodore and the early IBM and Apple desktop computers) the personal computer was at this point ready to invade private houses and to bring office applications and games into our daily life[13]. On the other extreme end of the spectrum the need to increase the computing power at the disposal of research groups had led to the introduction of massively parallel machines (such as the already mentioned *Connection Machine*) or to fancy workstations and computers tailored for efficient execution of high level languages (such as the *LISP Machine*).

These research directions had been boosted when, in 1982, Japan's Ministry of International Trade and Industry (MITI) announced the so-called *Fifth Generation Project* with the ambition to overtake the computer industry of the Western world by creating knowledge information processing systems based on an entirely new computer technology. The name derived from the consideration that vacuum tube computers had been the first generation, transistor machines the second, integrated circuit machines the third, and microprocessor based systems the fourth, while the fifth generation machines would be based on a revolutionary approach. The aim of MITI was to design machines based on massively parallel architectures oriented toward concurrent logic programming languages with the aim to address advanced artificial intelligence applications. The project was very ambitious. Logic would play a major role in the new systems under various aspects: to represent data and

---

[13] The first popular desktop computer, the IBM Personal Computer had been introduced in the market in 1981. A few years later a low level, very inexpensive computer, the Commodore 64, had been developed aiming at the broader consumer, and in January 1984 Steve Jobs presented the successful Apple Macintosh, finally introducing (on the basis of the previous Xerox Alto experience) the classical graphical user interface accompanied by the use of the mouse to which we became accustomed.

knowledge as logical axioms, to formalize the solution of problems in terms of logical goals, to realize computations as logical inferences. In particular two basic assumptions were that (i) the use of logic would enable and simplify the design of complex AI applications, and (ii) a specialized (non-von Neumann) architecture would allow very fast execution of logic programs[14]. Actually, ten years (and several hundred million dollars) later both assumptions turned out to be false. In particular the failure of the Fifth Generation Project was due to the fact that the highly parallel computer architectures specialized in the fast execution of logical inferences were eventually surpassed in speed by less specialized hardware.

In any case the big effort promoted by the Japanese Government had important consequences worldwide. First of all it was a strong stimulus toward the development of research in artificial intelligence. In those years, in fact, thanks to the most recent progress in the computing power of modern workstations, the field of artificial intelligence, in which the great expectations raised in the 1960s had been replaced with disillusionment and frustration (due to the insufficient computing power achievable in that period), became again a very popular research domain.

As we have already mentioned one of the most important fields in which artificial intelligence made its comeback was in the field of relational databases. The encounter between logic, artificial intelligence and relational databases, which had been inspired by the introduction of the logic based programming language PROLOG and by the idea to use a similar approach toward the management of data, had created, in the late 1970s and early 1980s new concepts (deductive databases, data and knowledge bases), new query languages (Datalog), and more generally the new field of knowledge representation.

In July 1982 the *European Coordinating Committee for Artificial Intelligence* (ECCAI, now the *European Association for Artificial Intelligence*, EurAI) was established with the aim to promote the study, research, and application of Artificial Intelligence in Europe. The first ECAI (*European Conference on Artificial Intelligence*) was held in Paris in the same year[15]. At the same time, most computer companies in Europe started to update their research programs including new activity in AI. In Italy this trend involved in 1983 the major software companies of that time: the Italsiel Group (through its hi-tech company Tecsiel), ENIDATA, Systems and Management, etc. In particular ENIDATA, a state-owned company, decided to push innovation in its activity, created a Scientific Advisory Board to identify and promote research projects, and established scientific collaborations with several universities. In this framework, in 1983 ENIDATA signed a research agreement

---

[14] In the language of the Fifth Generation Project the computer speed used to be measured in terms of millions of logical inferences per second instead of millions of instructions per second.

[15] Although the first in the series of ECAI conferences, for historical reasons the 1982 conference is called 5th ECAI.

with our department to develop joint research projects and, to this aim, they donated a LISP machine[16].

A similar reaction took place in various other countries, soliciting the definition of national research programs in information technology. In the UK, in 1982 the publication of the Alvey Report (prepared by a committee headed by John Alvey, head of R&D at British Telecom) led the Government to launch the *Alvey Programme*, a research program that ran from 1983 until 1987 and was focused on intelligent knowledge based systems, man-machine interaction, computer architectures for parallel processing, and VLSI.

Undoubtedly the most important consequence of the Japanese project came, in my opinion, at the European level, where the Japanese project (together with other factors) stimulated the creation and the launch in 1984 of the *European Strategic Program for Research and Development in Information Technology* (ESPRIT). In 1982 Etienne Davignon, at that time Commissioner for Industrial Affairs and Energy, had taken the initiative to promote a big joint European research effort that would address all the domains that were considered strategic for increasing the European competitiveness: energy, information technology, agriculture, raw materials, etc.

Finally Europe seemed to be ready to fight a battle to recover the technology gap that separated it from the US and avoid being overtaken by Japan. The First Framework Program was initiated in February 1984, with a budget of 3.75 billion euros. It was supposed to cover three years from 1984 to 1987. ESPRIT was part of the program, with a budget of 1500 million euros (750 funded by the EEC and 750 to be contributed by the participants in the selected projects). After a long preparation and discussions in the European Commission, carried on in collaboration with industries and universities, the weaknesses of European IT industry had been identified: "dispersion of R&D efforts, inadequate collaboration between university and industry, small dimension of markets and enterprises due to the fragmentation of national markets and to the current practices [by public administrations] to favour national solutions"[17]. According to the expectations of the European Commission ESPRIT would contribute to overcoming these deficiencies. The sectors addressed by the first ESPRIT program were somewhat similar to those considered in the Alvey program and included: microelectronics (VLSI design and computer-aided design systems), information processing systems (with great emphasis on computer networks and parallel computing), software technologies (formal methods and environments and tools for software engineering), advanced information processing systems (essentially artificial intelligence, logic programming and intelligent knowledge based systems), computer integrated manufacturing, and office automation.

---

[16] On the LISP machine, as I remember, a young and shy student, Umberto Nanni, now a professor in Rome, started to design a graphical interface for database queries as one of the Rome tasks in the DATAID project.

[17] ESPRIT European Strategic Research and Development Program on Information Technologies. Commission of the European Communities, CDNA11518ITC

The director of the first ESPRIT Program was Jean-Marie Cadiou, whose PhD thesis on fix-point semantics written in 1972 under the guidance of Zohar Manna we mentioned above. Cadiou was open to fundamental contributions and ESPRIT I accommodated several projects with a fundamental research character, generally carried out by consortia with both university and industrial partners. Several projects (such as for example GRASPIN, FOR-ME-TOO, etc.) were devoted to support environments for the use of formal methods in software development. Others were devoted to the use of formal description languages in the design of VLSI systems. Definitely the theoretical field in which the largest number of projects was selected was the field of applications of logic in knowledge and data bases. Among them: ALPE (Advanced Logic Programming Environments), coordinated by the University of Toulouse; EPSILON (Advanced Knowledge Base Management System), coordinated by the Italian company S&M, whose objective was to build an environment for the development and use of knowledge-based management systems (KBMS) based on standard technologies, namely relational databases and Prolog; ADKMS (Advanced Data and Knowledge Management Systems), coordinated by Siemens, whose objective was to develop a knowledge-based system with an inferential capability for the intelligent and efficient management of large databases, suitable for both naive and domain-expert users, and finally the project KIWI, coordinated by CRAI.

For my colleagues at CRAI the ESPRIT call had been exactly the right occasion to move one step forward in the direction that had been suggested by their president Sergio De Julio: to develop advanced basic research with a clear impact on future industrial applications. Mimmo Saccà started to work hard and to gather all his European contacts (including some of the top researchers in database theory in Europe) and in a few months the KIWI project (Knowledge-Based User-Friendly Interfaces for the Utilization of Information Bases) was ready. The project, in which were taking part the Dansk Datamatik Center, ENIDATA, INRIA, Philips Application & Software Services, the University of Calabria, the University of Antwerp, the University of Rome, was aimed at defining object-oriented languages for knowledge representation, integrating knowledge and databases and realizing graphical user interfaces to knowledge bases. The approval of the KIWI project was definitely a big success for a research center that was born just a few years before.

As we said above ESPRIT had been conceived with the aim to increase the competitiveness of the European computer industry. Whether or not this was the outcome is questionable, if we judge on the basis of what happened to some of what were in the early 1980s the major European computer companies, such as Bull, ICL, Olivetti. What is definitely sure is that the European research programs had instead a tremendous success in increasing the competitiveness of the European research in computer science both concerning systems, software and applications and concerning theory[18]. Before ESPRIT European research groups in theory of computing were usually isolated, in most cases their major contacts were in the US rather than in neighboring European countries, and, with a few exceptions, collaboration between academia and industry was absent. ESPRIT determined the creation of large networks of European scientists who, although living in different countries, collaborating with each other could eventually rival the major US academic centers. Besides, the research program was also beneficial in determining more strategic directions for European research in theoretical computer science. The need to define projects with an industrial fall-out forced many academic researchers to interrogate themselves on the meaningfulness of their activity and to focus their research work on more precise targets. The result of this overall effort was that the gap between US and European research in the theory of computing started to decrease[19].

A few years later, the launch of ESPRIT II and the provision in ESPRIT II of a special funding for basic research projects (the Basic Research Actions, BRA) brought further such positive results. First of all the amount of funds explicitly devoted to fundamental research (essentially corresponding to 65 million euros) allowed the universities to increase substantially the number of PhD students and of faculty positions in theory of computing. Besides, the coordinator of the BRA program, George Metakides (who had been educated as a logician), was aware of the fact that "supporting basic fundamental research in IT is a solid investment whose payback, even if it does not come in the form of short term industrial applications, will be great"[20], and he had a positive role in sustaining theoretical research in ESPRIT. Finally the fields addressed by the program were extended. While in ESPRIT I a major role had been played by logic in knowledge and data bases, in the 62 ESPRIT BRA projects (selected out of 283 submitted) several other theoretical domains were considered such as: semantics based program manipulation techniques (SEMANTIQUE), formal methods in programming, formal methods for VLSI, program correctness (PROCOS), theories of concurrency (CONCUR), natural language processing, robotics (FIRST), vision, neural networks (FULLY), human-computer interaction, machine learning, and graph rewriting (SEMAGRAPH). Of course logics continued to play an important role and a series of projects addressed knowledge bases, mechanized deduction in practical reasoning, reasoning and uncertainty (DRUMS),

---

[18] From the volume of the ESPRIT '85 Conference Status Report and Continuing Work it appears that already in the first phase of the program several European scientists involved in theoretical research contributed to the projects, among them, for example, Mogens Nielsen, Dines Bjørner, Marie-Claude Gaudel and the Italians Domenico Saccà, Beppe Attardi, and Elisa Bertino.

[19] To analyze the reason why, instead, on the technological side, the gap between Europe and the US in the computing field was not reduced in the last thirty years, but became rather wider, would go much beyond the scope of these notes.

[20] Ute and Wilfried Brauer, *The EATCS Silver Jubilee*, Bulletin of EATCS, 62 (June 1997).

logics for concurrent distributed systems, integration of logic, functional and object-oriented programming, and computational logic (COMPULOG). Last but not least, for the first time a project was devoted to strengthening the European research in algorithm design and complexity analysis (ALCOM[21]).

At the beginning of the 1990s theoretical computer science was a mature, well established and recognized discipline, part of the education of any computer scientist worldwide. Precisely in 1990 the *Handbook of Theoretical Computer Science* was published by Elsevier. The editor was Jan van Leeuwen (Fig. 10.5) and 48 scientists gave their contributions. The handbook consists of two volumes, *Volume A, Algorithms and Complexity*, and *Volume B, Formal Models and Semantics*, and it provides, in about 2000 pages, a rather complete picture of what were the basic chapters of the new science at that moment.

Fig. 10.5 Jan van Leeuwen

In Europe the research projects funded by the European Commission had contributed to the quantitative and qualitative growth of the field, and the fall of the Berlin Wall and of the Iron Courtain in 1989 had brought about the free circulation of ideas, researchers and scientific results between Eastern and Western Europe. In 1990 in Germany the first Max Planck institute

---

[21] The ESPRIT project ALCOM was one of the largest in terms of number of collaborating sites (12, belonging to 9 countries). It was coordinated by Jan van Leeuwen (Utrecht University) and thanks to its success in terms of scientific publications and of production of libraries of algorithms with a wide range of applications, it was followed by the projects ALCOM II, ALCOM-IT and ALCOM-FT. For more than a decade the ALCOM projects were the main support for the growth of European research groups working in the field. A similar role was played throughout the 1990s in the field of logic for computer science by the project COMPULOG (which was followed by a network of excellence with the same name) and, in the field of concurrent systems, by the project CONCUR (that gave rise to the well known CONCUR series of conferences).

entirely devoted to computer science was created: the 'Max Planck Institute for Informatics'. The institute was organized into two sections, Algorithms and Complexity, and Programming Logics, the directors being two of the most prominent German computer scientists, Kurt Mehlhorn and Harald Ganzinger respectively. In Italy, after the success of the first *Progetto Finalizzato Informatica* another project devoted to informatics was funded by the National Research Council: the *Progetto Finalizzato Sistemi Informatici e Calcolo Parallelo*, in which, as the title suggests, a major role was played by theoretical research in parallel computing.

In order to realize the worldwide impact of theoretical research in computer science in those years it can be observed that between 1988 and 1994 an impressive series of regional conferences on the theory of computing were started in all continents, among them: the *Scandinavian Workshop on Algorithm Theory* (SWAT), the *Israeli Symposium on Theory of Computing and Systems* (ISTCS), the (Canadian) *Workshop on Algorithms and Data Structures* (WADS), the *Latin American Theoretical Informatics Symposium* (LATIN), the (Asiatic) *Symposium on Algorithms and Computation* (ISAAC), the *Australasian Theory Symposium* (CATS), and the Italian (now International) *Conference on Algorithms and Complexity* (CIAC).

The field was by now covered by conferences that were becoming more and more specialized. Beside the classical broad spectrum conferences such as STOC, FOCS, ICALP, MFCS, and FCT and some of the older specialized ACM conferences (such as SOSP, POPL and SIGMOD, that used to present results of both practical and theoretical nature), a number of new specialized conferences, created during the 1980s and in the early 1990s, were devoted to the study of theoretical aspects regarding all domains of computer science. A non-exhaustive list: distributed computing (PODC, started in 1982), databases (PODS, started in 1982, ICDT, started in 1986), cryptology (CRYPTO, started in 1981, EUROCRYPT started in 1984), software design (TAPSOFT, started in 1985, ESOP, started in 1986, TACS, started in 1991, FME started in 1993), computational complexity (Structure in Complexity Theory, now CCC, started in 1986), computational learning (COLT, started in 1988, ALT, started in 1990), parallel computing (SPAA, started in 1989), algorithms (SODA, started in 1990, ESA, started in 1993), computational geometry (CG, started in 1991), and graph drawing (GD, started in 1992).

The important role in the field of informatics achieved in those years by theory was also made clear when, as we already mentioned, the *pro tempore* President of IFIP, Sendov, decided to create the Specialist Group on Foundations of Computer Science (SGFCS[22]). This acknowledgement, although it arrived too late, when IFIP was about to lose its international role, as a consequence of the changes in the world political scenario, namely the fall of the Iron Courtain, was, in any case, a great success for the theoretical

---

[22] In 1996 SGFCS was transformed by IFIP into a formal Technical Committee: TC 1 'Foundations of Computer Science', that I had the honour to chair from 1997 to 2003.

computer science community and for the scientist who had more than others struggled toward this goal, Józef Gruska.

Another event that testifies to the development of the domain is the increased visibility reached by EATCS at international level in those years. In 1985 Grzegorz Rozenberg had become president of EATCS while maintaining the role of Bulletin editor. Thanks to his extremely energetic and stimulating leadership the association undertook various initiatives. In various countries (France, Italy, Japan, etc.) national chapters were created, in some cases replacing pre-existing societies, and the Bulletin was given new appeal. Beside the classical reports from conferences and open problem sections, Rozenberg decided to include in the Bulletin 'columns' reporting recent results in a wide variety of fields (formal specification, formal languages, concurrency, distributed computing, algorithms, parallel computing, logic in computer science, computational complexity, etc.). The number of pages of the Bulletin boosted to over 500 and the membership of the association grew from a few hundred to over one thousand members from all continents (note that for a while the US members outnumbered all other nationalities, Germany being the first among the European countries).

But the event that most affected the very nature of the field of theoretical computer science around 1990, determining on one side a further widening of the themes addressed and on the other side a deep change in the philosophy and in the 'program', was the advent of the Internet. Let's see how Christos Papadimitriou, in a historical overview of 40 years of theory of computing given at ICALP in 2015, to celebrate the 40th anniversary of the journal Theoretical Computer Science, describes the situation at that stage. In the 1980s, he says, "we, the theoreticians, 'owned' the field of computer science in the sense that we knew how compilers should be designed, how operating systems should work, how databases should be organized for efficient access. Our responsibility and main mission was to outfit all areas of computer science with rigor and the power of mathematics. In a sense, we were exercising complete intellectual hegemony over the rest of computer science". We knew. We were teaching the students how the computing systems should improve, how applications should be made more effective. "Then the Internet happened". The Internet protocol for the interconnection of networks had been standardized in 1982 and since then various versions of transcontinental interconnection networks had been realized. At that point a deep transformation of the computing world started to materialize. In 1990 the ARPANET was decommissioned and the Internet started to evolve. In 1992 the World Wide Web was a reality.

After the advent of the Internet and of the Web for the first time a 'computational object' started to grow autonomously outside of our will. In a sense technology had taken the lead. We could only explain what was going on, we could design algorithms and models for the Web, we could exploit the huge computational power offered by the cloud but, somehow, we did not dominate the process anymore. While information technology was invading day by day the life of human beings, due to the increase in technological complexity and to the huge amount of data that flooded the planet, the role of theory was undergoing a profound transformation. Again in the words of Papadimitriou: "The Internet deeply changed computer science ... and theoretical computer science changed too: it moved toward math and opened up to the other sciences". Ideas from social sciences, natural sciences, economics, and game theory became necessary to understand and manage the new interconnected computing milieu. The role of theory and the work program for theoreticians had dramatically changed.

Appendices

Appendix A: Scientific Program of the First Symposium on Theory of Computing (STOC 1969)

Conference Record of

ACM SYMPOSIUM ON THEORY OF COMPUTING

Papers Presented at the Symposium
Marina del Rey, California
May 5, 6, 7, 1969

Sponsored by the

ASSOCIATION FOR COMPUTING MACHINERY

SPECIAL INTEREST COMMITTEE FOR AUTOMATA AND COMPUTABILITY

and

Co-Supported by

Systems Development Corporation

and

The University of Southern California

and

The University of Waterloo

Available from:

ACM
Order Department
1133 Avenue of the Americas
New York, N. Y. 10036

235

FOREWORD

The following papers will be presented at the ACM Symposium on Theory
of Computing, sponsored by the ACM Special Interest Committee for Automata
and Computability Theory. From the many papers submitted in response to
the Committee's call for papers, these were selected on the basis of
originality and relevance to the problems of current interest in the theory
of computing. However, the manuscripts have not been formally refereed and
some of the results represent a preliminary report of continuing research.
It is anticipated that most of these papers will appear in more polished
and complete form in scientific journals.

The Committee wants to thank all those who submitted papers for con-
sideration, the Systems Development Corporation and the University of
Southern California for their support of the Symposium, and the Editorial
Department of the ACM for their help in preparing this material.

Program Committee

Michael A. Harrison, Chairman
Robert W. Floyd
Juris Hartmanis
Richard M. Karp
Albert R. Meyer
Jeffrey D. Ullman

ORGANIZING COMMITTEE

Conference Chairman
Patrick C. Fischer
University of Waterloo
Waterloo, Canada

Local Arrangements Chairman
Seymour Ginsburg
University of Southern California
Los Angeles, California

Program Chairman
Michael A. Harrison
University of California
Berkeley, California

Coordinator
Ralph Layer
ACM Headquarters
New York, New York

SYMPOSIUM ON THEORY OF COMPUTING

Marina del Rey - May 5, 6, 7, 1969

TABLE OF CONTENTS

Page

TABLE OF CONTENTS

Page 2

<u>TABLE OF CONTENTS</u>

Page 3

Appendix B: First European Report on the Field of Theoretical Computer Science

2 2 JUIL. 1971

RAPPORT PRELIMINAIRE SUR L'INFORMATIQUE THEORIQUE

Dans les pages qui suivent nous tentons d'isoler et de délimiter aussi précisément que possible un domaine de recherche que, faute de mieux, nous appelerons Informatique Théorique.

Malgré cette dénomination il s'agit d'un domaine assez restreint qui ne prétend point recouvrir les nombreuses disciplines où des efforts de nature théorique se poursuivent depuis nombre d'années en liaison avec le développement des machines à calculer, et qui, même en comparaison avec ces disciplines, apparaît comme particulièrement peu développé tant par le faible nombre de chercheurs qui s'y adonnent que par la relative maigreur des résultats obtenus. Citons pour mémoire quelques unes de ces nombreuse disciplines qui depuis quelques années ont vu leur développement, même dans les directions les plus téhoriques, considérablement accéléré et amplifié par l'apparition et le perfectionnement des moyens modernes de calcul : analyse numérique, théorie du contrôle ou du signal, recherche opérationnelle, théori des graphes, intelligence artificielle etc ...

Si chacune de ces disciplines constitue, à bon droit, un chapitre de la vaste problématique que l'on désigne sous le nom d'Informatique, chacune d'elle fait plus usage de la notion de calcul qu'elle ne cherche à l'élucider ou à la préciser. Au contraire, c'est cette notion de calcul, dans ses aspects les plus variés et qui pour une large part échappe à la mathématique classiqu qui est au centre de ce que nous appelons Informatique Théorique. Le phéno- mène "calcul", manifesté par les innombrables machines à "calculer" qui ont envahi notre monde moderne appelle une analyse rigoureuse, de nature théorique, tout comme la physique a pu appeler jadis une physique théorique.

Mais il ne faut point s'étonner si depuis 15 ans l'informatique théoriqu ne s'est que peu développée : la prolifération des machines à calculer et l'extension formidable de leurs applications à la quasi-totalité des champs de la connaissance justifiaient pleinement que l'on s'interessât davantage dan un premier temps, à l'exploration des possibilités des machines plutot qu'a l'étude de leur nature profonde. Tout comme il est normal qu'un puissant regain d'intérêt se manifeste au contraire depuis deux ou trois ans pour les recherches théoriques concernant les machines et leurs langages. (Cet intérê est fort clairement exprimé dans les proceedings du colloque "Software Engineering" qui en 1968 regroupait un grand nombre de représentants des constructeurs sous l'égide de l'OTAN). Cela tient aux deux faits inséparables que d'une part la complication croissante des machines des langages et des programmes rend de plus en plus nécessaire le recours à des outils d'analyse mathématique ou logique, et que d'autre part d'incontestables et récents progrès ont été réalisés dans le domaine de l'Informatique Théorique.

Nous tentons donc ci-dessous de dresser un tableau et un bilan provisoire de l'Informatique théorique et surtout de délimiter plus précisém son domaine propre. Pour susciter ou favoriser le développement de cette discipline, qui, croyons-nous, conditionnera dans une large mesure le développement de toute l'Informatique à long terme, il importe de distinguer ce qui fait son originalité et son unité face aux nombreux autres champs de recherche, liés à l'Informatique, fussent-ils eux aussi très théoriques et de fondamentale importance.

On ne perdra pourtant pas de vue qu'il s'agit d'un domaine très ... ce qui entraine fatalement une part d'arbitraire dans le choix des problèmes et des résultats que nous décrivons ci-dessous comme étant les plus importants. Pour la même raison l'unité du sujet et le lien entre les trois chapitres fondamentaux que nous distinguons ci-dessous se situe encore bien plus au niveau de l'intuition et d'une commune façon d'aborder les divers problèmes, qu'a celui de principes ou de résultats suffisamment généraux qui seraient valables pour l'ensemble du domain en question.

I - ALGORITHMIQUE

Lorsque l'on fait de l'analyse numérique et plus généralement des mathématiques classiques on manipule des nombres entiers, réels ou autres qui sont définis abstraitement ainsi que les opérations de base que l'on peut effectuer sur eux; addition, multiplication etc ... Les algorithmes qui sont ainsi proposés sont indépendants de la représentation de ces nombres et l'analyse qui en est faite (rapidité de convergence etc ...) est la même, que la multiplication soit faite en virgule flottante, ou en utilisant le théorème du reste chinois . Pour l'informaticien au contraire les opérations élémentaires sont un inépuisable sujet d'étude. C'est qu'en effet la machine ne connait point de nombres sa mémoire formée d'éléments susceptibles de 2 ou plusieurs états et ordonnés séquentiellement ne représente naturellement que des suites de caractères c'est-à-dire des éléments d'un monoïde libre. Aussi avant de faire quelque opération que ce soit faut-t-il coder les nombres et, selon le codage retenu, ces opérations se révèlent plus ou moins faciles.

L'algorithme consiste pour nous en l'étude de la chaîne :

objet mathématique $\xrightarrow{\text{codage}}$ objet représenté $\xrightarrow{\text{opération}}$ objet représenté $\xrightarrow{\text{décodage}}$ objet mathématique

qui décrit la suite des opérations effectivement réalisées par une machi et dans laquelle les deux phases de codage et décodage sont essentielles Un résultat typique est celui qui affirme que quelque soit le mode de représentation choisi des deux entiers a et b comme suite de "digits" l'un (le digit préfixe) des digits de la suite représentant a + b est fonction de tous les digits figurant dans les représentations de a et b (ce résultat est dû à Sam Winograd). Ce point de vue dans l'étude des algorithmes peut se développer à divers niveaux, par exemple :

- les algorithmes fondamentaux portant sur les entiers seront analysés en fonction du mode de représentation de ces entiers ;

- les algorithmes portant sur des matrices à coéfficients entiers seront analysés en supposant parfaitement connues les opérations portant sur les entiers en fonction seulement du mode de rangement et de repérag des éléments de la matrice dans la mémoire.

Un des domaines privilégiés de l'algorithmique est l'algorithmiqu non numérique c'est-à-dire l'étude des algorithmes portant sur des objets qui ne sont pas des nombres (entiers, réels, complexes ou autres et qui occupent dans la mémoire de la machine une place variable. Il

peut s'agir de simples suites ordonnés (problème de tri) d'arbres
(problèmes de listes et fichiers), de graphes etc ... En effet les organes
de machine mis en cause par les problèmes d'algorithmique numérique
ou non numérique ne sont pas les mêmes. Les problèmes numériques
sont liés à la structure de l'unité centrale, qui repose sur le choix d'une
représentation, fait une fois pour toute - et que le programmeur ne peut
pas modifier - des objets fondamentaux manipulés. Les problèmes non
numériques posent surtout le problème de l'utilisation rationnelle de la
mémoire, quelle que soit la structure de celle-ci, et se reflètent donc
plus au niveau du "software" qu'à celui du "hardware". C'est en effet
généralement le langage de programmation utilisé, ou du moins son compilateur, qui spécifie le mode de représentation des données structurées
telles que chaînes, listes, tableaux etc ... dont l'étude fait l'objet de
l'algorithmique non numérique. Ou bien c'est au programmeur lui-même,
écrivant en langage d'assemblage ou même symbolique, de choisir le mode
de représentation adapté aux besoins particuliers du problème traité.

L'importance de l'algorithmique, surtout non numérique, apparaît
ainsi comme fondamentale dans la mesure où la définition et la compilation
de "bons" langages, et même de "bons" systèmes d'exploitation en dépendent
sans compter l'influence qu'elle peut avoir sur la conception d'organes
nouveaux au niveau du hardware. Et l'on ne peut plus laisser au hasard
de l'intuition le choix des structures que permet de manipuler un langage
de programmation destiné à un large usage.

Plusieurs remarques sont à faire ici :

- une branche de la logique porte le nom de théorie des algorithmes. Il
n'est pas question d'en minimiser ici l'intérêt. Toutefois l'algorithmique
que nous décrivons s'en distingue nettement dans la mesure où cette
théorie des algorithmes étudie essentiellement des fonctions de variables
entières à valeurs entières (et ramènent les autres à celles-ci par des
codages arithmétiques qui seraient, sur le plan pratique, désastreux) et
ne s'occupe pas de la représentation de ces entiers ;

- par contre apparaît comme cruciale en algorithmique la notion d'automate
les machines actuelles ne représentent, nous l'avons dit, et ne traitent
que des mots d'un monoïde libre. Or une application d'un monoïde libre
(ou d'un produit cartésien de monoïdes libres) dans un autre ne se décrit
pas, comme dans le cas des entiers, par une formule faisant intervenir
des opérations élémentaires : ces applications se décrivent naturellement
au moyen d'un automate qui n'est autre qu'un programme de calcul sim-
lifié et contraint. Plusieurs familles d'automates peuvent être utilisées
correspondant à des applications de complexité croissante. Les automate
finis, les plus simples, sont sans doute ceux qui intéressent le plus
l'algorithmicien, car nombre d'applications de base telle l'addition portar
sur des nombres entiers représentés par leur développement binaire sont
réalisables au moyen d'automates finis (de telles applications sont dites
transductions rationnelles). Des modèles réalistes de calcul sont ainsi
obtenus par assemblage de cellules élémentaires qui sont des automates
finis, et ce sont ces réseaux d'automates qui forment le cadre formel
dans lequel s'expriment le plus souvent les résultats actuels en algorithmic
qu'ils aient trait aux opérations arithmétiques ou au tri ou aux algorithme
de graphes. Nous reparlerons des automates dans la deuxième partie
de ce rapport.

Ayant situé ce que nous appelons l'algorithmique par rapport aux disciplines, parfois très proches, que sont l'analyse numérique ou la théorie des algorithmes que développent les logiciens, nous pouvons maintenant en dresser le bilan, d'ailleurs assez maigre :

- les opérations arithmétiques ont donné lieu à de nombreuses études. Nous avons signalé un résultat de S. Winograd sur l'addition. Des résultats ont été obtenus concernant la multiplication par exemple : si petit que soit ε il existe une constantes c (ε) telle que le nombre $T(n)$ d'opérations élémentaires nécessaires à la multiplication de deux entiers représentés par des suites de n digits satisfait

$$T(n) \leqslant c(\varepsilon) \ n^{1+\varepsilon}$$

C'est un résultat assez peu satisfaisant dans la mesure où l'algorithme en question se complique très vite quand ε décroit. Par contre l'arithmétique en multiple précision utilise des résultats, moins forts sans doute, mais pratiquement utilisables permettant de borner $T(n)$ par $n^{log_2 3}$ au lieu de n^2 que donne l'algorithme classique de multiplication. Le meilleur algorithme est celui de Toom-Cook qui conduit à $T(n) \leqslant n^{1+3,5/\sqrt{log_2 n}}$. Nous ignorons dans quelle mesure cet algorithme est effectivement utilisé. De même nature sont les problèmes concernant le calcul de x^n connaissant les deux entiers x et n (le minimum de multiplications nécessaires n'est pas connu) le calcul de la valeur d'un polynôme, les diverses opérations matricielles etc .. Dans plusieurs de ces cas des algorithmes ont été proposés qui améliorent sensiblement les algorithmes classiques dans le cas où le calcul doit se répéter un nombre assez grand de fois.

- les problèmes de tri sont peut-être ceux qui motivé le plus grand nomb de travaux d'algorithmique. Toutefois le nombre minimum d'"éléments comparateurs" dans un réseau qui sort la suite $a_1, \ldots, a_n$ rangée par ordre croissant des éléments, quand cette suite est entrée dans le rése n'est pas connu (Knuth et Floyd ont calculé ce minimum jusqu'à $n = 16$). Des algorithmes, issus de la théorie de l'information conduisent à un un nombre de comparaisons voisin de minimum théorique $\lfloor log_2 n! \rfloor + 1$: c'est au prix de nombreux calculs d'indices, coûteux en temps. A notr connaissance aucun résultat ne concerne la minimisation du coût global de l'algorithme c'est-à-dire de la somme des temps consacrés aux compa raisons, aux transpositions et au calcul des indices. Il faut d'ailleurs remarquer qu'une analyse fine devrait tenir compte non seulement de la machine utilisée mais du langage dans lequel est écrit le programme. L'absence ou la présence de registres d'index dans la machine, l'utilisation pour représenter la suite $a_1, \ldots, a_n$ d'entrée d'un tableau comm en FORTRAN ou ALGOL 60 plutôt que d'une file aux éléments repérés par des index, comme cela est possible dans des langages plus modern a certainement une influence sur l'algorithme optimum qui n'est pas le même dans tous les cas.

En fait, en dehors de la recherche, difficile, de l'optimum, la simple estimation du coût moyen, c'est-à-dire le calcul de l'espérance mathématique de la variable aléatoire "coût du rangement" de $a_1, \ldots,$ pose déjà de sérieux problèmes.

Signalons à ce sujet le 3ème volume de "The art of programming de D. Knuth, qui va paraître, entièrement consacré à ce sujet et très différent le livre de I. Flores sur le tri qui cherche à exprimer les algorithmes dans le langage d'Iverson ne se tenant au plus près des opérations élémentaires dont sont dotées les machines actuelles.

- les algorithmes portant sur les graphes ont été largement étudiés eux au et le problème du voyageur de commerce, toujours non résolu, a fascine maint chercheur (parmi eux M. Rabin ..) : il s'agit de trouver le nombr minimum d'opérations pour déterminer le chemin hamiltonien de plus faible coût dans un graphe-value complet.

Il importe de mentionner ici avant de clore le chapitre consacré à l'algorithmique le problème de la parallélisation, né de la structure modulaire des machines à calculer moderneset de la possibilité d'inter-connecter plusieurs unités de traitements. Il s'agit de délimiter dans un algorithme donné les phases qui, indépendantes les unes des autres, peuvent se dérouler simultanément jusqu'à ce qu'une phase ultérieure de l'algorithme interviennent qui utilisent effectivement les résultats de ces diverses phases traitées en parallèle. Il s'agit d'un problème extrêmeme difficile ou bien peu de résultats sont déjà acquis, et dont la définition même prête à controverse.

Egalement très liés à la structure des machines actuelles sont les problèmes de stratégies pour l'utilisation rationnelle d'une mémoire hiérarchisée ou fragmentée (allocation de mémoire, pagination etc ...). Ces problèmes sont les premièrs qui se posent à qui veut écrire un système d'exploitation, et leur importance ne saurait être surestimée. Les multiples solutions proposées, malgré leur intérêt ne nous paraissent pas constituer encore une véritable théorie dans la mesure ou les modèles de mémoire sous-jacentsont trop divers ou trop flous. Mais ce genre d'étude devrait se développer très rapidement dans un proche avenir

II - THEORIE DES AUTOMATES ET DES LANGAGES

Chose assez surprenante ce n'est pas l'informatique, fût-elle théorique, qui a été à l'origine de la théorie des automates. Les automate finis sont issus de modèles proposés dans les années 1950 pour décrire le fonctionnement du cerveau, et dans l'article fondamental de Kleene qui marque le début de la théorie(1956);ces automates sont encore présentés comme des réseaux de "neurones". Les automates à pile, liés aux langag "context-free" ont été d'abord considérés en détail par le linguiste Choms! pour faire l'analyse syntaxique des langues naturelles. Toutefois la théor: des automates et des langages (parties du monoîde libre susceptibles d'êtr analysées ou "reconnues" par automate) s'est vite développée de façon autonome en se dégageant de ses historiques avec la neurologie abstraite ou la linguistique, et par l'importance qu'elle revêt pour l'informatique, tôt reconnue, n'a cessé de s'affirmer.

Plusieurs types de problèmes sont à considérer :

1. le support de toute la théorie des automates et partant de la grande partie de l'informatique théorique est le monoîde libre, c'est-à-dire l'ensemble des mots écrits avec un certain alphabet. Il se trouve que, bien qu'il s'agisse d'un objet ma-thématique comme de tous les mathématiciens, il a fait l'obje

de peu de travaux et ses propriétés de nature combinatoire commencent à peine à être explorées. Il y a pourtant toute analyse un peu fine des langages. M. P. SCHUTZENBERGER a entrepris de façon assez systématique cette étude voilà longtemps et engagé nombre de ses élèves dans cette voie que jalousaient seulement un petit nombre de résultats dûs à des algébristes Malcev, Lyndon, Cohn etc ...

La théorie des équations dans le monoïde libre a été récemment développée par A. Lentin et se relie curieusement à celle du groupe symétrique.

Mais le problème le plus important est celui de la caractérisation des sous-monoïdes libres du monoïde libre, dont les ensembles minimisant de générateurs sont appelés codes. Le codage, nous l'avons souligné, est tout à fait essentiel à tous les niveaux du traitement de l'information. La théorie des codes se développe dans deux directions assez distinctes:

- les codes à longueur variable qui posent des problèmes plus théoriques que pratiques dans la mesure où il ne sont guère utilisés pour des codages effectifs (mentionnons quand même l'alphabet morse) mais dont la construction reste une des questions ouvertes majeures. Et leur étude, est celle des propriétés les plus intimes du monoïde libre ;

- les codes dits correcteurs d'erreur qui sont en fait des code à longueur fixe, dans lesquels une certaine redondance voulu permet de retrouver le message en dépit de quelques perturbations. Les problèmes que posent leur construction et leur étude sont cependant d'une nature tout à fait différente des précédents, faisant appel à d'autres domaines de l'algèbre, corps finis ou groupes simples. Ce domaine doit beaucoup aux travaux des chercheurs français.

Mention doit encore être faite ici de l'étude de nombreuses structures algébriques liées au monoïde libre et utiles dans toute la théorie des automates et des langages : algèbres associatives libres, monoïdes polycycliques, magmas ou magmoïdes libres etc ... A la limite il s'agit là de mathématiques tout à fait pures : il importe toutefois de souligner que toute une partie de l'algèbre, qui n'est pas la plus répandue ni la plus cultivée (et pratiquement pas enseignée en France), se révèle jouer un rôle essentiel dans le développement de l'Informatique Théorique. Toute augmentation de nos connaissances concernant l'une des structures libres mentionnées ci-dessus doit être considérée comme une contribution au domaine que nous tentons de définir.

2. Les automates finis sont ceux qui ont fait couler le plus d'encre Cela est dû à ce qu'ils fournissent un modèle qui rend compte non seulement de beaucoup d'opérations effectuées par les mach mais aussi des organes même de ces machines qui effectuent le dites opérations. Là encore on peut distinguer deux types de travaux :

-6-. /.

- l'étude algébrique des automates finis et des langages qu'ils recon-
naissent, dits langages réguliers, ou langages rationnels, est issue
des tous premiers travaux de Kleene qui ont fondé la théorie. Les
langages réguliers sont en effet susceptibles de deux définitions de
nature très différente, l'une combinatoire permettant d'exprimer un
tel langage comme une expression formée à partir d'ensembles finis
et d'opérations combinatoires, l'autre de nature algébrique permettan
d'associer à tout langage régulier un certain monoïde fini, dit monoïd
syntaxique. Les liens existant entre les propriétés combinatoires du
langage et la structure algébrique de son monoïde syntaxique sont très
difficiles à percer à jour et la théorie connaît plus de questions ouver
que d'autres. L'intérêt pourtant de ces recherches est considérable ;
s'y relient le problème de la décomposition des automates en élément
plus simples, c'est-à-dire le problème de la réalisation d'un automate fini par un montage en série ou en parallèle de petits automates
élémentaires. Khron et Rhodes ont établi un célèbre théorème qui
résout théoriquement ce problème et ouvre la voie à une théorie de
la complexité des automates finis et langages réguliers.

- à côté de ces problèmes algébriques, des problèmes pratiques liés
à la représentation physique des automates se sont posés dont la solut
n'est guère plus facile. C'est le cas notamment du "state assignemen
problem" qui consiste à réaliser de façon économique un automate fin
au moyen d'un assemblage de bascules, portes et relais. Il s'agit de
faire correspondre à chaque état de l'automate fini un certain "état"
du réseau de façon que la fonction de transition qui spécifie l'action
de l'automate soit après codage la plus simple possible. Bien que
ce soit un problème fini il n'existe pas actuellement d'algorithme
vraiment satisfaisant pour le résoudre. Des préoccupations d'ordre
pratique ont inspiré de nombreux autres travaux concernant notamment, la synthèse d'un automate dont on connaît le comportement les
automates incomplètement spécifiés etc ... Les informaticiens
soviétiques ont tout particulièrement contribué à ces recherches.

3. Automates à pile et langages algébriques. Le but de N. Chomsky qui
a défini les langages algébriques aux alentours de 1958 était de propose
un modèle intermédiaire de la grammaire des langues naturelles. Il s
trouve que c'est surtout pour la description de la syntaxe des langages
de programmation que les langages algébriques ont été utilisés, et ce de
la parution en 1960 du premier rapport Algol qui donnait de ce langage
une description qui n'est rien d'autre qu'une grammaire de Chomsky.
Par ailleurs M. P. SCHUTZENBERGER montrait, dès 1960 également
que les langages algébriques, en généralisant au cas non communicatif
des objets classiques en mathématiques, constituaient pour le mathénatic
un objet naturel d'étude. De très nombreux travaux ont été depuis dix
ans consacrés à ces langages et aux automates à pile qui leur correspo
dent comme les automates finis correspondent aux langages rationnels

- les plus nombreux de ces travaux concernent l'analyse syntaxique
c'est-à-dire la recherche des structures possibles d'une phrase du
langage au regard d'une grammaire qui définit ce langage. Cette
analyse est une phase obligatoire de la traduction de cette phrase
dans un autre langage, quel qu'il soit. C'est ainsi que tous les compi
lateurs de langages de programmation qui traduisent ceux-ci dans le

-7-. /.

langage de la machine comportent analyseur syntaxique qui n'est rien d'autre qu'un automate à pile ... (pour être tout à fait précis la plupart des langages de programmation sont donnés comme une partie d'un langage algébrique, partie que définissent un certain nombre de conditions aisément vérifiables quand on a fait l'analyse "algébrique" de la phrase donnée).

Une aussi intense activité est justifiée par le fait que les langages algébriques, dans toute leur généralité, ne se prêtent pas à une analyse syntaxique très rapide et la recherche s'est concentrée sur la définition de sous-classes de la classe des langages algébriques, sous-classes assez vastes pour englober la plupart des langages de programmation existants et cependant susceptibles d'une analyse syntaxique plus efficace. Il est impossible ici d'énumérer toutes les sous-classes qui ont été ainsi définies, et les trésors d'ingéniosité apportés à l'amélioration de leurs analyseurs syntaxiques. Disons seulement que de l'avis général il n'est pas possible d'entreprendre l'écriture d'un compilateur sans une bonne connaissance des techniques nombreuses d'analyse syntaxique qui sont nées de tous ces efforts, même si l'on ne peut se déclarer entièrement satisfait de leurs résultats. La multiplication des méthodes proposées et étudiées masque mal le fait que la structure des langages.algbriques nous est toujours mal connue.

- les études théoriques consacrées aux langages algébriques ont précisément pour but de percer à jour la structure de ces langages.- Si l'étude des langages rationnels se relie directeme à celle des monoïdes finis, objets aux-mêmes de nombreux travaux de la part de mathématiciens, celle des langages algé. briques conduit à la considération de structures mathématiques mal connues et peu étudiées telles que monoïdes polycycliques, congruences sur le monoïde libre etc ... Et c'est sans doute pour cela que les résultats concernant les langages algébriques sont relativement peu nombreux. Cette recherche théorique pourtant est très active tant aux Etats-Unis qu'en France : les efforts qu'elle motive conditionnent ` en grande partie le dével pement des langages de programmation et compilateurs puisque la théorie des langages algébriques fournit le cadre formel de tous les travaux concernant l'analyse syntaxique. De plus, l'emploi des séries algébriques, prolongement naturel des langages algébriques, renouvelle l'analyse combinatoire (ou théorie de l'énumération).

4. En dehors des automates finis et des langages rationnels, des automates à pile et des langages algébriques, de très nombreu autres familles d'automates et de langages ont été étudiées. S les travaux relatifs à la machine de Turing (qui est bien un automate) et à la notion de calculabilité que poursuivent de nombreux logiciens sont de nature sensiblement différente, il faut mentionner les études relatives aux "automates d'arbre" (en anglais "tree automata") qui présentent un intérêt certain pour la linguistique mais aussi pour la manipulation des donné structurées telles que listes et fichiers (qui ne sont autres que des arbres).

-8- ./.

III- SEMANTIQUE FORMELLE DES LANGAGES DE PROGRAMMATION :

La rapport Algol 68, comme le rapport Algol 60, se compose de paragraph alternativement intitulés syntaxe et sémantique. Dans les uns une grammaire est donnée qui définit certains types d'expressions, dans les autres il est dit, en style discursif, ce que sont ces expressions,
-et comment on les manipule. Depuis longtemps les progr meurs recherchent un moyen d'exprimer le contenu de ces paragraphes intitulés sémantique d'une façon qui ne prête pas, comme c'est actuellement le cas, à discussic ou interprétation. A tel point que l'IFIP a organisé en 1964 un colloque dont le titre était "langages pour la description des langages formels " et suscité depuis un groupe de travail international sur ce même sujet.

Il est évident que cette formalisation de la sémantique n'est pas un petit problème ne serait-ce que dans la mesure où, se proposant d'établir une correspondance précise entre certaines expressions d'un langage et certains objets mathématiques il faut d'abord s'assurer de la définition précise de ces derniers. Et chacun sait que la moindre notion mathématique ne se laisse pas définir si aisément ainsi que la logique mathématique est là pour l'attester. Longtemps on a voulu voir, comme seul moyen de définir de façon précise la sémantique d'un langage de programmation, la donnée d'un compilateur. C'était faire jouer à une machine déterminée un rôle particulier, et même, s'agissant d'une machine idéale supposée "simple", restait à définir la sémantique de cette machine. D'autre part un compilateur est chose complexe et cette façon de définir la sémantique rendait à peu près impossible d'aborder le problème de l'équivalence sémantique de programmes ou de la vérification de leur validité.

Des progrès décisifs ont été réalisés quand on a commencé à essayer de traduire les langages de programmation (il s'agissait d'algol 60) dans des systèmes formels, comme le lambda-calculus de Church, que les logiciens avaient édifiés pour préciser et étudier la notion de fonction. L'intérêt pour ce genre de recherche n'a fait depuis que croître, et de récents progrès permettent d'y voir un des plus féconds domaines de la recherche en Informatique Théorique. Les travaux récents dans cette direction suivent des voies assez différentes que nous essayons de décrire ci-dessous.

1. Un théorème de Floyd affirme que l'on peut associer à un programme une formule qui est une thèse du calcul des prédicats du second ordre si et seulement si - quand le calcul de ce programme pour certaines données initiales se termine - les valeurs des arguments à la fin du calcul satisfaisant un certain prédicat donné à l'avance. Cette formule en fait affirme l'existence d'un prédicat du 1er ordre, associé à chacun des tests du programme, exprimant les relations des divers arguments à ce point du calcul. Il y aurait là un moyen de vérifier qu'un programme effectue bien ce que l'on désirait qu'il fasse (et que traduit le prédicat donné), si l'on savait décider si une formule donnée est une thèse du calcul des prédicats du second ordre. En fait ce dernier problème est tout à fait insoluble dans le cas général. Par contre tous les travaux poursuivis en démonstration automatique peuvent être utilisés dans un certain nombre de cas où, connaissant à l'avance la forme des prédicats exprimant les propriétés des arguments en cours de calcul, la vérification se ramène à une problème du calcul des prédicats du premier ordre. De nombreux auteurs ont donc recherché à caractériser des classes de programmes pour lesquelles ce

phénomène de réduction se produit. Dans la même ligne,
à la suite de J. Mac Carthy, ont été étudiés · les schémas
d'induction et les types de démonstration pour récurrence
de propriétés des programmes et des programmes de
vérification automatique écrits qui produisent pour certaines
classes restreintes de programmes, des preuves de leur
validité. Bien qu'encore au stade expérimental, les applications à des cas concrets de diverses techniques de preuve,
ainsi développées, permettent d'entrevoir une utilisation
effective proche pour des programmes "simples", sans
procédure volontairement écrits en respectant un certain
nombre de règles concernant l'emboîtement des boucles,
ou la nature des tests etc ... (C'est le cas des programmes
de gestion).

2. La difficulté majeure à laquelle se heurtent tous les auteurs
soit de langages de programmation, soit de compilateurs est
celle qui naît de l'utilisation simultanée de variables simples
de divers types et de variables structurées telles que listes,
tableaux, files etc ... S'il est facile de décrire le mécanism
d'appel des procédures quand les arguments sont tous des
entiers (mécanisme qui n'est rien d'autre que la règle de
substitution pour le calcul des prédicats, au moins en Algol)
dès que les procédures peuvent avoir comme arguments des
variables structurées composées d'éléments de divers type
l'énoncé des règles régissant l'appel d'une telle procédure
(règles qui assurent que l'appel dans le contexte où il se trou
a bien un sens) devient extrêmement compliqué. En fait cet
énoncé suppose que la notion de fonction soit convenablement
élargie et que l'on dispose d'un moyen économique de noter
tant les fonctions que leurs arguments, même si ceux-ci son
des variables structurées. Le lambda-calculus de Church
et la logique combinatoire de Curry et Feys sont deux systèm
logiques, d'ailleurs très voisins, créés précisément dans ce
but. Ils nesauraient toutefois suffire tels qu'ils sont puisque
ni Church, ni Curry et Feys n'avaient résolu la question des
types des expressions, cette question étant tout à fait fonda
mentale si l'on veut utiliser ces deux systèmes formels pour
décrire la sémantique des procédures d'un langage de progra
mation, c'est-à-dire l'enrichissement de ces systèmes d'une
théorie convenable des types que travaillent indépendamment
Dana Scott (pour le lambda-calculus) et L. Nolin (pour la
logique combinatoire).

On touche là au coeur même des problèmes tant théoriqu
que pratiques posés par la programmation sur le plan théori
C'est surtout en effet par l'élargissement de la notion traditi
nelle de fonction que la programmation à contribué à enrich
le champ des mathématiques. Sur le plan pratique il va de s
qu'une description formelle, précise, des mécanismes liés
à l'utilisation des procédures, serait un outil fondamental de
progrès dans l'utilisation des langages (et bientôt sans doute
des systèmes qui ne font que composer des "fonctions" de ty
non classique, certes).

-10-./.

3. Plus à ras de terre se situent un autre genre de travaux, à savoir ceux qui proposent une axiomatique directe des langages de programmation. Il s'agit là de donner des axiomes et des règles d'inférence telles que deux programmes équivalents puissent être transformés l'un dans l'autre par application répétée de ces axiomes et de ces règles. Une axiomatique complète des instructions d'affectation ou des instructions de saut est possible mais bien évidemment le problème général de l'équivalence des programmes ne peut être traité de cette façon. Il y a pourtant là un moyen de mettre oeuvre directement sur les programmes tels qu'ils sont écrits certaines techniques de simplification ou de rationalisation qui pourraient pratiquement se révéler utiles. Cette démarche, qui consiste à réduire à des problèmes syntaxiques (équivalence de mots d'un modèle) le problème sémantique de l'équivalence de deux programmes (dits équivalents quand ils calculent la même fonction) a également donné naissance à la théorie des schémas de programme. Pour éviter que l'équivalence sémantique de deux programmes, résulte du fait que ces deux programmes calculent la même fonction par deux méthodes mathématiquement différente (et en ce cas, la preuve de l'équivalence est bien un problème de mathématique et non un problème d'informatique) on considère dans cette théorie les opérations et prédicats de base comme dépourvus de toute signification. Un schéma de programme représente ainsi toute une classe de programmes dont les élémen sont obtenus en interprétant les symboles de base. Posé en ces termes, le problème de l'équivalence de deux schémas, équivalen et seulement si toute interprétation commune aux deux schémas c duit à deux programmes sémantiquement équivalents au sens précé dent, devient un problème purement informatique. Et cette approche a le mérite de ne pas faire intervenir l'arithmétique.

Ancienne (Ianov 1952), puis longtemps en sommeil, la théorie des schémas se développe à nouveau rapidement (Paterson, Lukham, Strong etc ...). Un problème central est celui de la traduction des schémas récursifs en schéma non récursifs. Contrairement à ce qui se passe dans un système contenant l'arithmé tique, cette traduction n'est pas toujours possible. Les recherche dans cette direction apportent des vues tout à fait nouvelles sur ce qui est encore un des problèmes essentiels de la compilation (suppression de la récursivité).

Avant de clôre ce chapitre il convient de parler ici de la théorie des systèmes. Il s'agit des systèmes d'exploitation dont l'écriture et la mise au point est certainement un des problèmes pratiques les plus importants de toute l'Informatique. Un tel système est un programme qui régit les rapports des diver composantes de la machine (unité (s) centrale (s), unités de mémoires, périphériques etc ...) gère aussi bien le temps que l'espace du calcul. Il va sans dire qu'un tel programme est d'une extrême complexité comparée surtout à celle des algorithmes ou programmes qu'étudient actuellement l'algorithmique ou la séman tique formelle des langages de programmation. Cette différence de degré de complexité, beaucoup plus que de nature, est telle qu' nous paraît difficile de voir appliquer avec succès dans un proche avenir les méthodes et les résultats dont nous avons parlé à l'étud

-11-.

des systèmes. Et ce malgré un grand nombre d'efforts en ce ser
(citons par exemple Djikska, ou Wirth et Naur). Il n'en reste
pas moins que, ainsi que nous l'avons a plusieurs endroits
souligné , l'étude des systèmes doit devenir, s'il n'est déjà ,
le lieu privilégié d'application de l'Informatique Théorique, à la
fois source d'inspiration et référence pour juger de l'intérêt des
modèles du calcul qui constituent son domaine propre.

CONCLUSION

Il ne nous semble pas avoir oublié dans les pages précédentes de
chapitres importants de l'Informatique Théorique mais il ne nous a pas ét
possible de mentionner de nombreuses recherches de détail, n'ayant donn
lieu qu'à une ou deux publications. Nous sommes conscients d'ailleurs qu
l'Informatique Théorique est dans son ensemble une discipline trop récent
pour que l'on puisse en dresser un tableau exhaustif et séduisant. En fait
notre propos était surtout d'effectuer un choix de problèmes qui, tout en
étant de nature fort différente, aient en commun de contribuer à l'édifica-
tion d'une théorie du calcul réaliste qui constituerait un modèle convenabl
des processus actuellement effectués par les machines. Nous y avons
rajouté les problèmes liés à l'étude des structures et objets mathématique
sous-jacents, étude sans laquelle notre compréhension des phénomènes dι
calcul resterait limitée et superficielle. Nous croyons que déjà l'on peut
voir se dessiner les contours d'une problématique et la possibilité d'unifiε
les efforts jusqu'à présent un peu désordonés des chercheurs qui depuis
dix ans ont contribué à l'Informatique Théorique.

Appendix C: First Proposal to Create a European Association for Theoretical Computer Science

UNIVERSITÉ DE PARIS VII

DÉPARTEMENT
de MATHÉMATIQUES

Tour 45-55
9, quai Saint-Bernard
75 - PARIS V
TÉL. 336-25-25 - Poste 37-61

Paris, le 24 juin 1971

Monsieur le Président,

Nous confirmons volontiers ce qui avait déjà été indiqué verbalement dans le cadre de votre enquête officieuse, à savoir que : l'Université Paris VII est prête à engager une coopération avec des Universités d'états membres de la C.E.E. dans le domaine de l'enseignement de l'Informatique et dans celui de la recherche en Informatique théorique (cette recherche nous paraissant particulièrement importante de par son incidence sur l'enseignement).

Si vous en êtes d'accord, nous prendrons l'initiative des contacts sur ce sujet avec les Universités suivantes : pour l'Allemagne, Université de Sarrebruck; technische universität de Munich. Pour l'Italie, Université de Rome, Université de Pise, Université de Turin ; pour les Pays-Bas, Université d'Amsterdam ; pour la Belgique Université libre de Bruxelles. Pour la France, nous prendrions des contacts avec les Universités Paris VIII et Toulouse que nous croyions susceptibles de se joindre à Paris VII dans une telle coopération.

Des contacts pourraient être facilement pris également avec les Universités anglaises suivantes : Warwick, Edimbourg et Colchester.

Les objectifs d'une telle coopération en matière d'enseignement pourraient être :

1° - d'harmoniser à terme et dans toute la mesure du possible des programmes d'enseignement de l'Informatique dans les différentes Universités concernées.

- 2 -

2° - de mettre sur pied d'éventuels cycles d'études se déroulant en périodes
    successives dans plusieurs Universités.

3° - de favoriser autant que possible la circulation des étudiants d'une
    Université à l'autre.

4° - de favoriser les échanges d'enseignants (séjour de six mois ou d'un an).

5° - de mettre sur pied des sessions de recyclage d'enseignants sur des sujets
    spécialisés.

Dans le domaine de la recherche, une telle coopération aurait pour
objet :

- de favoriser au maximum les échanges d'informations entre les différentes
  équipes de recherche dans le domaine de l'Informatique théorique, ainsi
  que la diffusion des résultats obtenus.

- d'organiser à cet effet des stages de jeunes chercheurs dans des équipes
  différentes de leur équipe d'origine et des colloques spécialisés auxquels
  pourraient prendre part des chercheurs débutants.

Les modalités juridiques de la coopération devraient être à notre sens
les plus souples possibles et prendre la forme d'un simple accord inter-univer-
sitaire.

Pour notre Université Paris VII, les contacts seront conduits Par
Messieurs NIVAT, NOLIN et SCHUTZENBERGER qui sont Professeurs à l'U.E.R. de
Mathématiques. Je vous prie d'agréer, Monsieur le Président, à l'expression de
mes sentiments les plus distingués.

François BRUHAT
Directeur de l'U.E.R. de Mathématiques

Appendix D: Original Statutes of EATCS

ASSOCIATION EUROPEENNE D'INFORMATIQUE THEORIQUE

STATUTS

1. Ausiello, Giorgio, chercheur, Via Vigevano, 6, Rome, Italie,

2. de Bakker, Jacobus, chercheur, Van Breestraat, 147, Amsterdam, Pays-Bas,

3. Nivat, Maurice, professeur, rue Portalis, 9, Paris, France,

4. Paterson, Michael, professeur, Kenilworth Road, 108, Coventry, Grande-Bretagne

5. Paul, Manfred, professeur, Eggmühlerstrasse, 4, Munich, Allemagne Fédérale,

6. Sintzoff, Michel, chercheur, avenue du Château, 7, Rixensart, Belgique,

7. Verbeek, Leo, professeur, Nieuwe Plantagestraat, 5, Delft, Pays-Bas,

ont décidé de constituer, par acté sous seing privé, une association internationale à but scientifique dont les statuts sont libellés comme suit :

TITRE Ier - Dénomination, Siège

Article 1er. Il est constitué, suivant la loi belge du 25 octobre 1919, modifiée par la loi du 6 décembre 1954, une association internationale à but scientifique et appelée "Association Européenne d'Informatique Théorique", dont le siège est situé à Rixensart, avenue du Château, 7.

TITRE II - Objet

Article 2. L'Association a pour objet de promouvoir le développement de la recherche et de l'enseignement en informatique théorique en Europe.

Elle remplira cet objet, notamment :

a) en établissant, entre ses membres, des échanges d'étudiants, d'enseignants et de chercheurs;

b) en organisant des réunions, séminaires et conférences;

c) en diffusant des résultats de recherches;

- 2 -

d) en organisant, entre ses membres, des recherches et des enseignements en commun;

e) en coordonnant les recherches et les enseignements de ses membres;

f) en organisant des recherches et des enseignements avec des personnes ou institutions qui ne sont pas membres.

TITRE III - Membres

Article 3. Les membres de l'Association sont :

a) des personnes morales, dénommées ci-après <u>instituts associés;</u>

b) des personnes physiques, dénommées ci-après <u>membres individuels.</u>

Article 4. L'Association est composée initialement par les soussignés. En outre, deviendront membres de l'Association les personnes morales ou physiques invitées par le Conseil et déclarant leur volonté d'adhérer au Conseil.

Article 5. La qualité de membre implique l'adhésion sans réserve aux présents statuts et au règlement.

Article 6. La qualité de membre se perd :

a) par la démission, qui pourra être donnée moyennant préavis de six mois au moins, pour la fin de l'année civile;

b) par l'exclusion, qui doit être décidée par une assemblée générale.

Tout membre qui cesse de faire partie de l'Association est sans droit sur l'actif social.

TITRE IV - Organes

Article 7. Les organes de l'Association sont :

l'Assemblée Générale et

le Conseil.

A. <u>L'Assemblée Générale</u>

Article 8. Une Assemblée Générale Ordinaire des membres sera convoquée tous les trois ans à l'endroit fixé par le Conseil et à partir de l'année 1973 durant laquelle, à titre exceptionnel, se tiendra une Assemblée Générale Ordinaire.

Un membre peut en représenter un autre ou deux autres mais pas davantage.

- 3 -

Article 9. La réunion des membres en Assemblée Générale Extraordinaire
a lieu sur décision du Conseil ou sur demande écrite d'un cinquième des
membres lesquels doivent indiquer avec précision l'ordre du jour ou les
questions dont ils proposent la discussion.

Article 10. L'Assemblée Générale est l'organe suprême de l'Association;
elle a tous les pouvoirs.

Article 11.

1. Toute Assemblée Générale ne statue valablement que si la moitié des
   membres individuels et la moitié des instituts associés sont présents
   ou représentés. Les votes sont acquis à la majorité simple des membres
   individuels et à la majorité simple des instituts associés, présents
   ou représentés.

2. L'ordre du jour des Assemblées Générales est arrêté par le Conseil.
   Il comprend toute question proposée par un membre sauf décision moti-
   vée du Conseil refusant l'inscription, ainsi que toute question dont
   l'inscription est demandée par au moins un cinquième des membres.

B. Conseil de l'Association

Article 12.

1. L'Association est dirigée et administrée par un Conseil composé de
   sept à quinze membres élus par l'Assemblée Générale.

2. Le Conseil désigne parmi ses membres un bureau qui se compose :

   a) d'un président, lequel est également président de l'Association;

   b) de deux vice-présidents;

   c) d'un secrétaire général;

   d) d'un trésorier.

   Le bureau comprend au moins un membre de nationalité belge.

Article 13. La durée des fonctions des membres élus au Conseil est de
trois ans à compter de 1973, le mandat des administrateurs élus pour la
première fois venant à échéance à la date de l'Assemblée Générale Ordinaire de 1973.

   Le président est rééligible une fois au plus.

Article 14.

1. Le Conseil est chargé de la gestion de l'Association. Il prend toutes
   les décisions qui ne sont pas réservées à l'Assemblée Générale.

2. Le secrétaire général assure la gestion courante. Il nomme et révoque
   le personnel de secrétariat.

- 4 -

Article 15. L'Association est engagée par la signature de son président ou, en cas d'empêchement de celui-ci, d'un vice-président. Le Conseil pourra en outre déléguer certains de ses pouvoirs à l'un de ses membres ou à un secrétaire administratif. L'Association est représentée en justice par son président ou par un administrateur délégué à cet effet par le Conseil.

TITRE V - Dissolution et modification

Article 16.

1. La dissolution de l'Association ou la modification des présents statuts ne peut être décidée que par un vote réunissant au moins les deux tiers des membres individuels et les deux tiers des instituts associés, présents ou représentés. En cas de dissolution, l'Assemblée Générale désigne un ou deux liquidateurs et détermine leurs pouvoirs.

2. L'adhésion de l'Association à toute organisation visant certains buts de l'Association pourra être décidée par une Assemblée Générale dans les conditions prévues à l'article 11.

Article 17. En cas de dissolution, le solde actif du patrimoine social sera, sur décision du Conseil, versé à une association scientifique analogue.

ASSOCIATION EUROPEENNE D'INFORMATIQUE THEORIQUE

Composition du Conseil :

   a) président : L. Verbeek

   b) vice-présidents: M. Paterson, M. Paul

   c) secrétaire général : M. Nivat

   d) trésorier : M. Sintzoff

   e) autres membres : J. W. de Bakker, G. Ausiello

Date de constitution :   24 juin 1972

Appendix E: Invitation to Join the Newborn EATCS

UNIVERSITÉ PARIS VII
U. E. R. DE MATHÉMATIQUES
———————
Tour 45-55 - 5me Etage
2, Place Jussieu
75 - PARIS 5me
Tél. 336-25-25   Poste 37-61

Dear Colleague,

   This letter is an invitation to join the young European Association
for Theoretical Computer Science. Let me begin with some history : in
January 1972 a small meeting was held in the building of the European
Communities Commission ; M. Caracciolo chaired a discussion between a
few computer scientists from six countries known for their theoretical
work. First, a text written by the members of the University of Paris VII
who were present at the meeting was presented and discussed ; this text
aimed at the definition of what can be called Theoretical Computer Science,
i.e. the chapter of Computer Sciences which uses mathematical and logical tools to clarify and study the notion of computation. The principal
subchapters would be :

   − Theory of algorithms and complexity

   − Theory of automata and formal languages

   − Theory of programming (formal semantics)

but this list is certainly not complete.

   At the same time the creation of a European Association was proposed, to promote the development of this field by organizing meetings,
exchanges of researchers, professors and students, and by increasing the
communication of all sorts of information (it is rather difficult to
know what happens in a field which changes completely in one year if one
considers that two years are necessary for the publication of a paper in
a journal). An original feature of this association would be to accept
both individual members, known scientists working in the field, and

institutional members, University departments, research laboratories or
any other organisation interested in the development of Theoretical
Computer Science.

It appeared that the Belgian law on scientific association allows
such a mixture of two kinds of members and more generally offers very
liberal and flexible possibilitées. This is why, after some months of
work, essentially done by our first temporary President M. Sintzoff,
statutes were written according to this law and the Belgian authorities
were asked to create such an association. It was necessary that some
people should act as founding members and share the first responsabili-
ties. These people are :

L. Verbeek (Président), M. Paterson and M. Paul (Vice-Presidents),
M. Sintzoff (Treasurer), M. Nivat (Secretary), J. de Bakker and G.
Ausiello.

Obviously these people were designated only by chance and each of
them represents only himself. And no less obviously, this association
will play the role we wish it to play, in the development of the field,
only if a large part of the scientists working in this field become
members.

It is therefore very important that you agree to increase the
membership of this association and that, together with the founding members, you not only work to make the association viable but also help in
the definition of its scope, aims, composition and means of action. The
temporary officers of the association look forward to your active participation.

I cannot indeed tell you much about future activities, which will
depend on you and all the people who receive this letter. M. Paterson
has generously offered the University of Warwick as the location for a
necessary General Assembly, and at the same time an informal colloquim
which will be the first scientific activity of the association. This
will take place in March 1973. We have also discussed the idea of a journal or periodic letter to convey information among members (and non-mem-
bers !). Suggestions concerning this and other actions will be most

3.-

welcome, in particular an early priority is to assure the financing of
our activities.

The enclosed copy of the statutes will help you to get a better
idea of what we are looking for and, I hope, will give you strong
reasons to join our common effort.

Yours sincerely,

M. NIVAT

APPEL AUX COMMUNICATIONS

Ce colloque international sur la Théorie des Automates, des Langages et de la Programmation organisé par l'IRIA sous le patronage de l'ACM Special Interest Committee for Automata and Computability Theory (SIGACT), abordera les sujets suivants :

• **Théorie des Automates et Langages Formels**

• **Théorie des Algorithmes et de la Complexité**

• **Formalisation des Notions de Machines et de Programme**

• **Sémantique des Langages de Programmation**

COMMUNICATIONS

Toute personne souhaitant présenter une communication sur l'un des thèmes du colloque est invitée à compléter et à retourner immédiatement la carte réponse ci-jointe puis à soumettre au Comité du Programme un résumé de 4 ou 5 pages dactylographiées avant le 15 mars 1972.

LANGUES

Les langues officielles seront l'allemand, l'anglais et le français. Il n'y aura pas de traduction simultanée.

INSTRUCTIONS AUX AUTEURS

Sur la première page du résumé les indications suivantes devront être portées :
Titre − nom de l'auteur(s) − du co-auteur(s) − fonction − ville − pays − adresse et téléphone.

Les auteurs sauront le 20 avril au plus tard si leur texte a été accepté ou non.

Le texte final (maximum 20 pages) devra être remis au secrétariat du colloque au cours des journées.

Théorie des Automates
des Langages
et de la Programmation

3-7 juillet 1972

RIA INSTITUT DE RECHERCHE
D'INFORMATIQUE ET D'AUTOMATIQUE
Domaine de Voluceau
78 - ROCQUENCOURT - FRANCE

PROGRAMME

Le programme complet du colloque sera diffusé le 15 mai 1972.

ADRESSE

Comité du Programme et d'Organisation
IRIA - Service des Relations Extérieures
Domaine de Voluceau
B.P. 5 - 78 - Le Chesnay - France
Téléphone : 954.90.20 - Poste 600
Télex : IRIA 60.033 F

LIEU DU COLLOQUE

IRIA - Domaine de Voluceau - Rocquencourt France

COMITÉ DU PROGRAMME	
Président	
M. Schützenberger	IRIA et Université Paris 7
Membres	
C. Boehm	Université de Turin - Italie
S. Eilenberg	Université de Columbia New York - Etats Unis
P. Fischer	Université de Waterloo - Canada
S. Ginsburg	Université Southern California, Los Angeles - Etats Unis
G. Hotz	Université de Sarrebruck - Rép. Féd. d'Allemagne
M. Nivat	Université Paris 7 et IRIA - France
L. Nolin	Université Paris 7 - France
D. Park	Université de Warwick - Grande Bretagne
M. Rabin	Université de Jérusalem - Israël
A. Salomaa	Université de Turku - Finlande
A. van Wijngaarden	Mathematische Centrum, Amsterdam - Pays Bas

Appendix G: Scientific Program of the First ICALP
(1972)

COLLOQUES IRIA

Théorie des automates
des langages et de la programmation
3-7 juillet 1972

résumés des communications

TABLE DES MATIERES

Appendix H: Call for Papers of the First MFCS (1972)

Computation Centre
Polish Academy of Sciences
Warsaw, PKiN P.O. Box 22
Poland

Institute of Computing Machines
Warsaw University
Warsaw, PKiN
Poland

FIRST COMMUNICATION

The Computation Centre of the Polish Academy of Sciences together with the Institute of Computing Machines of the Warsaw University are organizing an international symposium and summer school devoted to:

MATHEMATICAL FOUNDATIONS OF COMPUTER SCIENCE
WARSAW, AUGUST 21—27, 1972

Because of a large scope of this subject it is not possible to deal with all its branches during a one-week meeting. For this reason the organizers intend to limit this scope to problems that are developed in Poland, i.e. to semantics of programming languages and foundations of computing systems. In particular the following topics will be dealt with: proving properties of programs /correctness, equivalence, adequacy, etc./ and related topics, program transformations, primitives of programming languages and computing systems.

The program of the summer school will contain several invited lectures presenting polish results in the field.

The symposium will have short addresses, contributed papers and sessions devoted to clarifying open problems.

The meeting will take place in Jabłonna - an XVIII-th century palace near Warsaw /15 km and a good bus service/. There will be time and space for informal talks and discussions. Accommodation for foreign participants and invited speakers will be available in the palace.

The financial means which the Organizing Committee has at its disposal cover the expences of about 10 invited persons. The Committee hopes that all the Colleagues interested in the subject of the meeting will be able to attend as delegates of their own institutions and extends to them cordial invitation. Of course the Committee will help all the participants with visa formalities, accomodation and other necessary facilities.

The registration fee is the equivalent of 20 USA dollars payable in the currency of the participant's country. Details concerning the bank account to which the fee should be paid will be given in the second communication. Delegates from socialistic countries pay 480 polish złoty after arrival at the conference.

Persons who wish to receive further information about the meeting are requested to fill the encolosed form and mail it at the address:

Dr hab. Andrzej Blikle
Computation Centre, Polish Academy of Sciences
Warsaw, PKiN P.O. Box 22. Poland
marking the envelope "SUMMER SCHOOL".

ORGANIZING COMMITTEE

/Mrs/ Prof. H. Rasiowa /Chairman/
Prof. Z. Pawlak /vice Chairman/
Prof. S. Turski /vice Chairman/
Dr A. Blikle /Secretary/
Dr P. Dembiński

Dr A. Mazurkiewicz
Dr E. Orłowska
Dr A. Salwicki
Dr A. Waligórski
Dr J. Wińkowski

Appendix I: Minutes of the First EATCS General Assembly

March 29, 1973
BHM/km

European Association for Computer Science Theory

The first general assembly of the association was held at the Univeristy of Warwick on March 24-25, 1973. The principle points of discussion on the first day were:

> the aims of the association,
> what it could do to further these aims,
> its relations to existing computer societies,
> how far membership should be open,
> the election of a council.

There was some divergence of opinion, but the discussion did produce a measure of agreement on the following points:

> circulation of an informal news bulletin,
> support of informal working conferences (typically 25 invited
>   specialists in a particular topic)
> organisation of regular formal conferences
>   (like that at IRIA last year)
> membership conditions (see enclosed application form)

The general assembly elected the following members of the council:

Ausiello,	Nivat
Böhm	Paterson
Brauer	Perrot
De Bakker	Scott
Mayoh	Sintzoff
Milner	Ver Beek

For reasons of flexibility (relations with Eastern European countries and the like) 3 council places were left vacant.

The council had a short meeting on the second day devoted primarily to the election of executive officers:

Nivat	(president)
De Bakker	(vice president)
Paterson	(vice president)
Sintzoff	(treasurer)
Mayoh	(secretary)

Organisational details were also discussed. Contributions to the first issue of the news bulletin are most welcome and should be sent <u>as soon as possible</u> to either Nivat or Mayoh. The issue should appear before July 1.

Brian Mayoh

Appendix J: First Proposal to Create the Journal Theoretical Computer Science (Agreed Between Maurice Nivat and Einar Fredriksson)

UNIVERSITE DE
PARIS VI - U.E.R. 110

INSTITUT DE PROGRAMMATION

TOUR 55-65 - 11, QUAI SAINT-BERNARD - 75-PARIS V°
TEL. 336.25.25 OU 325.12.21 - POSTE : 53.90

Paris, 18th June 1973

Dear Colleague,

I send you a copy of a proposal from North Holland Publishing Company regarding the creation of a new journal of Theoretical Computer Science. This proposal was made after discussions I had with Mr. E. Fredriksson. in which we asked ourselves wether it would be possible to have our European Association play a role in this creation.

The idea we came to was that the Association could appoint say 5 members of the future Editorial Board among whom hopefully one would be the Managing editor and that these 5 people would discuss with North Holland the numerous questions to be solved : other members of the editorial board, scope of the journal etc...

An interesting idea was to make this journal easily available in Easten European Countries by having it also sponsored by some official organism from one of these countries.

May I ask you to reply promptly about what you think of this proposal
    - opportunity of a new journal
    - possibility of the European Association to take part in this creation, by some procedure close to the one proposed by Mr. Fredriksson.
    - names of people you would like to see as a member of the editorial board or as the managing editor.

Mr. Fredriksson insisted on keeping this matter rather confidential until our Association has defined its policy and eventually reached an agreement with North Holland. I think that on such an important question we need wait until a next meeting of the Council (very likely in Hamburg in October) to come to a decision but I woul like to inform E. Fredriksson of your first reactions.

I'll inform you of other matters of interest for our Association in a next letter.

Yours sincerely

M. NIVAT

Theoretical Computer Science

An International Journal

As publishers in mathematics and computer science, North
Holland Publishing Company is interested to publish a new
periodical with the proposed above title. Specifically,
we propose the following:

1.   The aim of this journal is to bring together research
papers, possibly also surveys, short communications and
reviews of recent publications in the field of theoretical
computer science.

2.   The scientific guidance of the journal is undertaken
by an international editorial board, of which e.g. five
members are appointed by the newly formed "European Association
of Theoretical Computer Science."

3.   The editorial staff of the journal consists of a
board of editors, a subset of which is elected as advisory
editors and a managing editor.

4.   The tasks of the editors are to promote the publication
of valuable results, and to referee, or have refereed, papers
in their fields of interest. The advisory editors assist the
managing editor in making policy decisions.

5.   The managing editor is elected by the board of advisory
editors, after consultation with the publisher, for a term of
e.g. four years.

6.   The board of advisory editors may make changes in the
editorial board as needed. Moreover, it will automatically
take over the responsibilities of the managing editor for
the duration of the latter's term, in case of disability.

7.   The managing editor will be responsible for the final
acceptance of contributions to be published in the journal.
The publisher will remuniate the managing editor by an
honorarium and also cover expenses for his secretarial
assistance. This will be done on a per volume basis.

8.   The journal is initially scheduled to appear in
an annual volume of four quarterly issues. A volume will
consist of about 400 printed pages with approximately 400
words to a page, formulae, figures and tables included. As
date of publication of the first issue, March 1974 is
suggested.

9.   Manuscripts to be submitted to the journal
should be typed in the English language. In exceptional
cases papers in French may be accepted. The papers will
be subject to refereeing and the processing of the
manuscripts will be as follows:

(i) manuscripts are to be sent by the authors to
the managing editor;

(ii) the managing editor will distribute the papers
to the most appropriate editor who will have them refereed;

(iii)the manuscripts are to be returned to the
managing editor by the editor who has had them refereed
together with the referees' reports and his personal
recommendations. The referees' reports should be written
in such a way that they could be sent to the authors. In
principle, all refereeing will be done anonymously and the
names of the referees will only be disclosed to the authors
with the consent of the referees.

10.   The complete manuscript for an issue of the journal
must reach the publisher six months before the intended
publication for the issue date. Special invitations for
contributions to the first volume will be discussed
between editors and publisher. (The selection of papers
for the first volume is especially important as it gives
indications of which directions the editors wish the
journal to take.)

11.   Members of the editorial board will receive a
free subscription to the journal for the duration of
their active cooperation.

12.   40 reprints free of charge will be made available
for each contribution, while further reprints can be
ordered.

Appendix K: First Issue of the EATCS Bulletin Edited by Maurice Nivat

association européenne d'informatique théorique

european association for theoretical computer science

BULLETIN N° 1 décembre 1973

INSTITUT
DE RECHERCHE
D'INFORMATIQUE
ET D'AUTOMATIQUE
DOMAINE DE VOLUCEAU, ROCQUENCOURT-BP 5
78150-LE CHESNAY : 954.90.20-TELEX : IRIA 60033F

- 1 -

EDITORIAL

     The European Association for Theorectical Computer Science aims mainly
at increasing the communication between the various european universities laboratories
and research centers where there are people at work in this area. To achieve this
the Association will publish a bulletin, of which this is the first issue, sponsor
or organize meetings and help in all possible ways the exchanges of researches,
students or professors between these centers.

     Obviously a non-profit association, the EATCS was oreated according
to belgian low in September 1972. For the time being it has no other resources
than the very small fee asked to members. And it relies entirely on the good
will of every body who is interested in the aims of the association. This is
especially true of this bulletin. The french Institut de Recherche en Informatique
et Automatique (IRIA) prints and distributes it gratuitiously. The content will
be all the informations that people may wish to spread among european but also
american theoretical computer scientist. I say american too for very likely
there will be an exchange made between SIGACT and EATCS to make the SIGACT news
available to members of EATCS and our bulletin available to SIGACT members. We
thus ask every body who reads these lines to send us material for the next issue
of the bulletin as soon as possible.: this material, in any language, should
only be typed in a reproduceable form. We would like too to be able to publish
research announcements and may be short notes similar to those which one can
find in the SIGACT news.

     The association has now a small membership of around fifty but we are
confident that researchers in our field will all join in a near future. We enclose
an application form in this issue of the bulletin. Please return it to the treasurer
Dr M. SINTZOFF, with or without signature of two members.

     The main planned activities besides the publication of a bulletin is
the organization and sponsorship of meetings :

1) most of you have noticed that EATCS sponsors the "second international
   symposium on Automata, Languages and Programming" organized by our german
   colleagues in Saarbrucken in July 1974.

- 2 -

2) EATCS also sponsors an advanced course on foundations of computer science to be
held in Amsterdam in May 1974 and will sponsor, we hope, some other conferences
in 1974.

3) We shall take decisions in the next council meeting on the organization of a
working conference in 1975 : this conference will likely take place in Italy
and be devoted to programming theory. Further details will be given in the next
issue of the Bulletin.

Let us know about the meetings which are planned, any where in Europe,
so that we can at least announce them.

A last word on the scope of Theoretical Computer Science : a general
agreement was reached to consider as the major ponts of Theoretical Computer
Science the following fields of interest

- Theory of Automata
- Formal Languages
- Algorithms and their complexity
- Theory of programming, if such a thing exists : we may now consider as chapters
  of this theory formal semantics, proving properties of programs, and all
  applications of logical concepts and methods to the study and design of
  programming languages.

This list is by no means limitative : defining the limits of Theoretical
Computer Science is at least as difficult than defining the limits of Computer
Science itself. And we strongly believe that a science is what the scientists at
work make it : certainly new areas of computer science will be open to theory
in a very near future. Let us start small and grow : we are sure many of you will
help to achieve this necessary growth up to the point where our Association will
be a natural link between all European Theoretical Computer Scientists.

Bonne Année à tous

M.NIVAT
Président de l'EATCS

Appendix L: First Issue of the Journal Theoretical Computer Science (June 1975, Reprinted on the Occasion of the 30th Anniversary of the Journal)

EDITOR-IN-CHIEF:

M. NIVAT

ASSOCIATE EDITOR:

M. S. PATERSON

EDITORIAL BOARD:

J. W. DE BAKKER
J. BEČVÁŘ
C. BÖHM
R. BOOK
E. ENGELER
E. ERSHOV
M. HARRISON
S. IGARASHI
R. KARP
A. MEYER
R. MILNER
D. PARK
Z. PAWLAK
M. RABIN
A. SALOMAA
A. SCHÖNHAGE
J. ULLMAN

Volume 1
Number 1   June 1975

theoretical computer science

1 (1) 1-94 (1975)

CONTENTS

30th Anniversary

NORTH-HOLLAND PUBLISHING COMPANY · AMSTERDAM

Theoretical Computer Science

Editor-in-Chief
M. NIVAT

Associate Editor
M. S. PATERSON

VOLUME 1

Appendix M: Second Issue of the EATCS Bulletin
(First with the Classical Red Cover; Edited by Giorgio
Ausiello)

EUROPEAN ASSOCIATION
FOR
THEORETICAL COMPUTER SCIENCE

ASSOCIATION EUROPÉENNE D'INFORMATIQUE THEORIQUE

Bulletin n° 2 — DECEMBER 1976

C O N T E N T S

1

CHAIRMAN'S LETTER

The European Association for Theoretical Computer
Science has been created four years ago. Roughly, the
aim of the creators was to provide scientists in this
area with a structure comparable to that of SIGACT in
the USA: organization of meetings, colloquia, conferences...
and publication of a bulletin containing specific news,
announcements and also scientific notes or open problems.
We felt such a thing was even more necessary in Europe
than in the United States for people know little of each
other even when working 400 Km apart: which researcher
from Paris can tell anything of what happens in Leyden,
or Colchester or Darmstadt. And how often do we go to
fetch the answer to some question as far as California
when this answer could be given by a colleague in a
nearby University?

A difference with SIGACT, which turned out to be a
major one, is that we are not linked with any other associa
tion, like ACM, and thus cannot find the support SIGACT
initially found. The association exists as a Belgian
association with a special international status that is
authorized by the Belgian law. But a difficulty arose
immediately which is that the organisms in France for
example which could have given some financial support
to a French association were unable to give it to a
Belgian one! At the beginning we were expecting some
funds from the ECC, but these funds never came despite
the favorable opinion of an international board of experts.

The situation is thus clear: we have no money and can
only expect to have the money we can collect from the
members. However, thanks to individual efforts, the
association has held three international conferences
"Automation, Programming and Languages" (Paris 1972,
Saarbrucken 1974, Edinburgh 1976) and other conferences
in the same series are planned in Turku (1977) and some
where in Northern Italy (1978). The association gave its
sponsorship to some smaller conferences too, such as the
conference on λ-calculus in Rome 1975.

As concerns the bulletin our efforts were less successful,
a single issue was printed in IRIA and ill distributed,
and it appeared that relying on the good will of this or

2

that organism like IRIA cannot ensure any regularity in
the publication and distribution of such a bulletin.

That is why a new attempt in now made: Giorgio Ausiello
proposed himself to edit the bulletin in Rome. You can
see, from the present issue, what it looks like and we
hope you are pleased with it.

What we need to continue this experience is:

- first: papers to fill the bulletin; do not hesitate,
  send a note on the activity around you, send announce
  ments of reports or meetings, send the small theorem
  or amusing remark you have in your drawer lying use-
  lessly.

- second: money. We have money for two issues only.
  We need, in order to publish this bulletin on a
  regular basis, roughly 200 people paying each 5 $
  or so a year. Thus if you wish this bulletin to
  live, please either send money or find new members
  for the EATCS among your friends and colleagues.
  An application form is enclosed in this issue.

For the future of the association we believe this
bulletin is essential: if we can go on, the EATCS will
come to adult life. If not we shall remain an informal
gathering of friends, almost unknown and always fragile.
There will be a general assembly in Turku at the occasion
of the international conference: the present council and
officers of EATCS will be replaced or reelected, and the
policy for the next years defined. I hope you will all
like to participate in this unusual and fascinating venture
which is the life of a truly international, independent
professional association whose only aim is to promote
good research in a challenging field of science.

M Nivat

Appendix N: First Announcement of the Cetraro Seminar on Theoretical Issues in Data Bases (1981)

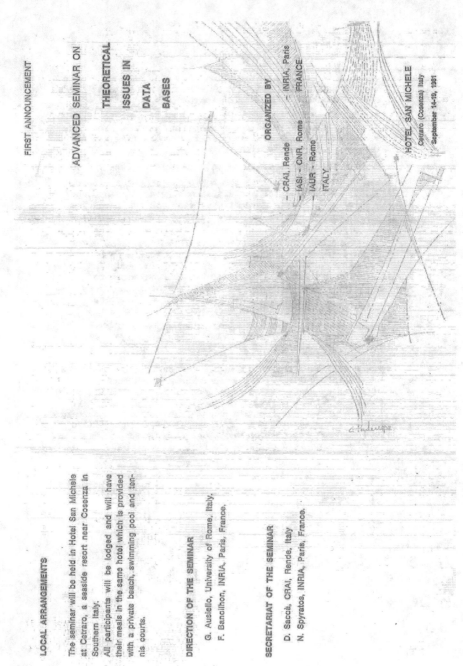

FIRST ANNOUNCEMENT

ADVANCED SEMINAR ON

THEORETICAL ISSUES IN DATA BASES

ORGANIZED BY

– INRIA, Paris FRANCE

– CRAI, Rende
– IASI - CNR, Rome
– IAUR - Rome
ITALY

HOTEL SAN MICHELE
Cetraro (Cosenza) Italy
September 14-19, 1981

LOCAL ARRANGEMENTS

The seminar will be held in Hotel San Michele at Cetraro, a seaside resort near Cosenza in Southern Italy.

All participants will be lodged and will have their meals in the same hotel which is provided with a private beach, swimming pool and tennis courts.

DIRECTION OF THE SEMINAR

G. Ausiello, University of Rome, Italy.
F. Bancilhon, INRIA, Paris, France.

SECRETARIAT OF THE SEMINAR

D. Saccà, CRAI, Rende, Italy
N. Spyratos, INRIA, Paris, France.

In connection with the Fifth International Conference on Very Large Data Bases the

— Institut National de Recherche en Informatique et Automatique, Paris, France
— Consorzio Calabro per la Ricerca e le Applicazioni nel Settore dell'Informatica, Rende (CS) Italy
— Istituto di Analisi dei Sistemi e Informatica del CNR, Roma, Italy
— Istituto di Automatica dell'Università di Roma, Italy

organise the First International Advanced Seminar on Theoretical Issues in Data Bases.

PURPOSE OF THE SEMINAR

In the recent years data bases have become a major area of interest in Computer Science. As the applications of data bases become more sophisticated, the Data Base Management Systems become more complex and the end-user's interfaces become increasingly concerned with the conceptual content of data more than with the representational details. So, the need for a more comprehensive study of the theoretical aspects of data bases is widely recognized. Of course relevant results have been achieved and have a significant impact on the design of advanced Data Base Management Systems. However there remain several unresolved issues.
The aim of the seminar is hence twofold:

i) to survey the state of the art in data base theory,

ii) to evaluate the perspectives for a formal approach to the solution of unresolved problems.

MAIN TOPICS AND LECTURERS

The main topics that will be covered during the lectures will be the following:

A. Data Base Design (normalization, view design, equivalence, etc.)
   G. Ausiello, University of Rome, Italy
   F. Bancilhon, INRIA, France
   G. Bracchi, Polytechnic of Milan, Italy

B. Data Dependencies
   C. Beeri, the Hebrew University, Israel
   R. Fagin, IBM San José, USA
   Y. Sagiv, University of Illinois, USA

C. Formal treatment of semantic aspects (semantic data models, integrity constraints, incomplete information, etc.)
   R. Kowalski, Imperial College, UK
   J. Mylopoulos, Univ. of Toronto, Canada
   J. M. Nicolas, ONERA-CERT, France
   C. Zaniolo, Bell Laboratories, USA

D. Theory of concurrency and consistency in data bases
   E. Gelenbe, INRIA, France
   C. Papadimitriou, MIT, USA

E. Impact of software design methodologies on data base design (abstract data types, etc.)
   H. Weber, University of Bremen, Germany

This is not intended to be an exhaustive list of topics.
The final program will be included in the second announcement.

SCIENTIFIC PROGRAM

During the seminar, beside the lectures given by the speakers, that will be detailed in the final announcement, short presentations of research results will be given by participants and invited to the workshop.
Besides, two panels will be organized on issues related to the Italian CNR Project in Computer Science:

— methodologies and tools for data base design;
— distributed data base management systems.

REGISTRATION

The seminar fee is Lit. 500,000 plus V.A.T. (15%). The fee includes the registration, the documentation and full board and lodging from Sunday September 13 to Sunday September 20.
The payment should be made by cheque payable to CRAI, via Modigliani, 87036 Roges di Rende (CS), Italy.

PARTICIPATION

The number of participants will be limited approximately to 50.
Application for attendance at the Seminar must be received on or before July 30.
A limited number of scholarships covering (fully or partially) the registration and full board and lodging will be granted to researchers on the base of a justification of need for financial help and of a scientific curriculum.

References

1. ABITEBOUL, S., HULL, R., AND VIANU, V., Eds. *Foundations of Databases: The Logical Level*, 1st ed. Addison-Wesley Longman Publishing, Boston, MA, USA, 1995.
2. AHO, A. V., GAREY, M. R., AND ULLMAN, J. D. The transitive reduction of a directed graph. *SIAM Journal on Computing 1*, 2 (1972), 131–137.
3. AHO, A. V., AND HOPCROFT, J. E. *The Design and Analysis of Computer Algorithms*. Addison-Wesley Longman Publishing, Boston, MA, USA, 1974.
4. APOSTOLICO, A. The myriad virtues of subword trees. In *Combinatorial Algorithms on Words*. Springer, 1985, pp. 85–96.
5. APOSTOLICO, A., AND PREPARATA, F. P. Optimal off-line detection of repetitions in a string. *Theoretical Computer Science 22*, 3 (1983), 297–315.
6. ARMSTRONG, W. W. Dependency structures of data base relationships. In *IFIP Congress 74* (1974), pp. 580–583.
7. ARORA, S., LUND, C., MOTWANI, R., SUDAN, M., AND SZEGEDY, M. Proof verification and the hardness of approximation problems. *Journal of the ACM 45*, 3 (1998), 501–555.
8. AUSIELLO, G. *Complessità di calcolo delle funzioni*. Boringhieri, 1975.
9. AUSIELLO, G., D'ATRI, A., AND PROTASI, M. On the structure of combinatorial problems and structure preserving reductions. In *Proceedings of the 4th Colloquium on Automata, Languages and Programming* (1977), pp. 45–60.
10. AUSIELLO, G., D'ATRI, A., AND SACCÀ, D. Graph algorithms for functional dependency manipulation. *Journal of the ACM 30*, 4 (1983), 752–766.
11. AUSIELLO, G., D'ATRI, A., AND SACCÀ, D. Minimal representation of directed hypergraphs. *SIAM Journal on Computing 15*, 2 (1986), 418–431.
12. AUSIELLO, G., PROTASI, M., MARCHETTI-SPACCAMELA, A., GAMBOSI, G., CRESCENZI, P., AND KANN, V. *Complexity and Approximation: Combinatorial Optimization Problems and Their Approximability Properties*. Springer, 1999.
13. BACKUS, J. W. Can programming be liberated from the von Neumann style?: a functional style and its algebra of programs. *Communications of the ACM 21*, 8 (1978), 613–641.
14. BACKUS, J. W., BAUER, F. L., GREEN, J., KATZ, C., MCCARTHY, J., NAUR, P., PERLIS, A., RUTISHAUSER, H., SAMELSON, K., VAUQUOIS, B., ET AL. Report on the algorithmic language ALGOL 60. *Numerische Mathematik 2*, 1 (1960), 106–136.
15. BAR-YEHUDA, R., AND EVEN, S. A local-ratio theorem for approximating the weighted vertex cover problem. *North-Holland Mathematics Studies 109* (1985), 27–45.
16. BATINI, C., AND D'ATRI, A. Rewriting systems as a tool for relational data base design. In *Proceedings of the International Workshop on Graph-Grammars and Their Application to Computer Science and Biology* (1979), pp. 139–154.

© Springer International Publishing AG, part of Springer Nature 2018
G. Ausiello, *The Making of a New Science*,
https://doi.org/10.1007/978-3-319-62680-2

17. BATINI, C., AND SANTUCCI, G. Top-down design in the entity-relationship model. In *Proceedings of the 1st International Conference on the Entity-Relationship Approach to Systems Analysis and Design* (1980), pp. 323–338.

18. BAUER, F. L. *Origins and Foundations of Computing*. Springer, 2010.

19. BAYER, R. Symmetric binary B-trees: Data structure and maintenance algorithms. *Acta Informatica 1*, 4 (1972), 290–306.

20. BEERI, C., BERNSTEIN, P. A., AND GOODMAN, N. A sophisticate's introduction to database normalization theory. In *Proceedings of the 4th International Conference on Very Large Data Bases* (1978), pp. 113–124.

21. BEERI, C., FAGIN, R., MAIER, D., MENDELZON, A., ULLMAN, J., AND YANNAKAKIS, M. Properties of acyclic database schemes. In *Proceedings of the 13th Annual ACM Symposium on Theory of Computing* (1981), pp. 355–362.

22. BEERI, C., FAGIN, R., MAIER, D., AND YANNAKAKIS, M. On the desirability of acyclic database schemes. *Journal of the ACM 30*, 3 (1983), 479–513.

23. BEN-OR, M., GOLDREICH, O., GOLDWASSER, S., HÅSTAD, J., KILIAN, J., MICALI, S., AND ROGAWAY, P. Everything provable is provable in zero-knowledge. In *Proceedings of the 8th Annual International Cryptology Conference on Advances in Cryptology* (1990), pp. 37–56.

24. BERGE, C. *Théorie des graphes et ses applications*. Collection universitaire de mathématiques. Dunod, Paris, 1958.

25. BERKELEY, E. C., AND BOBROW, D. G. *The Programming Language LISP: its Operation and Applications*. Information International Incorporated, Boston, 1964.

26. BERRY, G. Stable models of typed lambda-calculi. In *Proceedings of the 5th Colloquium on Automata, Languages and Programming* (1978), pp. 72–89.

27. BLUM, L., AND BLUM, M. Toward a mathematical theory of inductive inference. *Information and Control 28*, 2 (June 1975), 125–155.

28. BLUM, M. A machine-independent theory of the complexity of recursive functions. *Journal of the ACM 14*, 2 (1967), 322–336.

29. BLUM, M., AND MICALI, S. How to generate cryptographically strong sequences of pseudo-random bits. *SIAM Journal on Computing 13*, 4 (1984), 850–864.

30. BÖHM, C. Calculatrices digitales. Du déchiffrage de formules logico-mathématiques par la machine même dans la conception du programme. *Annali di Matematica Pura ed Applicata 37*, 1 (1954), 175–217.

31. BÖHM, C. Gleaning the past and the future in computer science. *Bulletin-European Association for Theoretical Computer Science 74* (2001), 178–189.

32. BÖHM, C., AND JACOPINI, G. Flow diagrams, Turing machines and languages with only two formation rules. *Communications of the ACM 9*, 5 (1966), 366–371.

33. BORODIN, A. Juris Hartmanis: Building a department—Building a discipline. In *Complexity Theory Retrospective*. Springer, 1990, pp. 19–27.

34. BRAUER, U., AND BRAUER, W. Silver jubilee of EATCS. *Bulletin-European Association for Theoretical Computer Science 62* (1997), 3–23.

35. BURSTALL, R. Program proving as hand simulation with a little induction. In *IFIP Congress 74* (1974), pp. 308–312.

36. BURSTALL, R. Recursive programs: proof, transformation and synthesis. *Rivista di Informatica VIII*, 1 (1977).

37. CALUDE, C. S. *People & Ideas in Theoretical Computer Science*. Springer, 1999.

38. CHAITIN, G. J. On the simplicity and speed of programs for computing infinite sets of natural numbers. *Journal of the ACM 16*, 3 (July 1969), 407–422.

39. CHEN, P. P.-S. The entity-relationship model – toward a unified view of data. *ACM Transactions on Database Systems 1*, 1 (1976), 9–36.

40. CHOMSKY, N. Three models for the description of language. *IRE Transactions on Information Theory 2*, 3 (1956), 113–124.

41. CHOMSKY, N. On certain formal properties of grammars. *Information and Control 2*, 2 (1959), 137–167.

42. CHOMSKY, N., AND MILLER, G. A. Introduction to the formal analysis of natural languages. *Journal of Symbolic Logic 33*, 2 (1968), 299–300.

43. CHRISTOFIDES, N. Worst-case analysis of a new heuristic for the travelling salesman problem. Tech. Rep., DTIC Document, 1976.

44. CODD, E. F. A relational model of data for large shared data banks. *Communications of the ACM 13*, 6 (1970), 377–387.

45. CODD, E. F. Further normalization of the data base relational model. *IBM Research Report, San Jose, California RJ909* (1971).

46. CODD, E. F. Relational completeness of data base sublanguages. *IBM Research Report RJ987* (1972).

47. CODD, E. F. Recent investigations in relational data base systems. In *IFIP Congress 74* (1974), pp. 1017–1021.

48. CODD, E. F. Derivability, redundancy and consistency of relations stored in large data banks. *SIGMOD Record 38*, 1 (June 2009), 17–36.

49. COFFMAN JR, E. G. J., GAREY, M. R., AND JOHNSON, D. S. Approximation algorithms for bin-packing – an updated survey. In *Algorithm Design for Computer System Design*. Springer, 1984, pp. 49–106.

50. COOK, S. A. The complexity of theorem proving procedures. In *Proceedings of the 3rd Annual ACM Symposium on Theory of Computing* (1971), pp. 151–158.

51. CRESCENZI, P., AND PANCONESI, A. Completeness in approximation classes. *Information and Computation 93*, 2 (1991), 241–262.

52. CURIEN, P.-L. Une brève biographie scientifique de Maurice Nivat. *Theoretical Computer Science 281*, 1 (2002), 3–23.

53. DAHL, O.-J., DIJKSTRA, E. W., AND HOARE, C. A. R. *Structured Programming*. Academic Press, 1972.

54. DATE, C. J. *Logic and Databases: The Roots of Relational Theory*. Trafford Publishing, 2007.

55. DAVIS, M. *Computability and Unsolvability*. McGraw-Hill, New York, 1958.

56. DI BATTISTA, G., EADES, P., TAMASSIA, R., AND TOLLIS, I. G. *Graph Drawing: Algorithms for the Visualization of Graphs*. Prentice-Hall, Upper Saddle River, NJ, USA, 1999.

57. DIFFIE, W., AND HELLMAN, M. New directions in cryptography. *IEEE Transactions on Information Theory 22*, 6 (1976), 644–654.

58. DIJKSTRA, E. W. Algol 60 translation: An Algol 60 translator for the X1 and making a translator for Algol 60. Tech. Rep. 35, Mathematisch Centrum, Amsterdam, 1961.

59. EDMONDS, J. Paths, trees, and flowers. *Canadian Journal of Mathematics 17*, 3 (1965), 449–467.

60. EHRIG, H., AND MAHR, B. *Fundamentals of Algebraic Specification I*. Springer, 1985.

61. EHRIG, H., AND MAHR, B. *Fundamentals of Algebraic Specification 2: Module Specifications and Constraints*. Springer, 1990.

62. FEIGE, U., GOLDWASSER, S., LOVÁSZ, L., SAFRA, S., AND SZEGEDY, M. Approximating clique is almost NP-complete (preliminary version). In *Proceedings of the 32nd Annual Symposium on Foundations of Computer Science* (1991), pp. 2–12.

63. FITCH, F. B. Representation of sequential circuits in combinatory logic. *Philosophy of Science 25*, 4 (1958), 263–279.

64. FLYNN, M. J. Some computer organizations and their effectiveness. *IEEE Transactions on Computers 21*, 9 (1972), 948–960.

65. GALLAIRE, H., MINKER, J., AND NICOLAS, J. *Advances in Data Base Theory, vol. 2*. Plenum Publ. Corp., 1984.

66. GAREY, M. R., AND JOHNSON, D. S. *Computers and Intractability; A Guide to the Theory of NP-Completeness*. W. H. Freeman & Co., New York, NY, USA, 1990.

67. GAREY, M. R., JOHNSON, D. S., PREPARATA, F. P., AND TARJAN, R. E. Triangulating a simple polygon. *Information Processing Letters 7*, 4 (1978), 175–179.

68. GOGUEN, J. A., THATCHER, J. W., WAGNER, E. G., WRIGHT, J. B., ET AL. Abstract data types as initial algebras and the correctness of data representations. *Computer Graphics, Pattern Recognition and Data Structure* (1975), 89–93.

69. GOLD, E. M. Language identification in the limit. *Information and Control 10* (1967), 447–474.

70. GOLDREICH, O., MICALI, S., AND WIGDERSON, A. Proofs that yield nothing but their validity or all languages in NP have zero-knowledge proof systems. *Journal of the ACM 38*, 3 (1991), 690–728.

71. GOLDWASSER, S., MICALI, S., AND RACKOFF, C. The knowledge complexity of interactive proof-systems. In *Proceedings of the 17th Annual ACM Symposium on Theory of Computing* (1985), pp. 291–304.

72. GOTLIEB, C. C., AND BORODIN, A. *Social Issues in Computing*. Academic Press, 2014.

73. GRIES, D. *Compiler Construction for Digital Computers*. John Wiley & Sons, New York, NY, 1971.

74. HANSEN, P. B. *Operating Systems Principles*. Prentice-Hall, Englewood Cliffs, NJ, 1973.

75. HARTMANIS, J., AND STEARNS, R. E. On the computational complexity of algorithms. *Transactions of the American Mathematical Society 117* (1965), 285–306.

76. HOPCROFT, J., PAUL, W., AND VALIANT, L. On time versus space. *Journal of the ACM 24*, 2 (Apr. 1977), 332–337.

77. HOPCROFT, J. E. An $n \log n$ algorithm for minimizing the states in a finite-automaton. In *Theory of Machines and Computations*, Z. Kohavi, Ed. Academic Press, NY, 1971, pp. 189–196.

78. HOPCROFT, J. E., AND ULLMAN, J. D. *Formal Languages and Their Relation to Automata*. Addison-Wesley, Reading, 1969.

79. IANOV, Y. I. The logical schemes of algorithms. In *Problems of Cybernetics*, vol. 1. Pergamon Press, New York, 1960, pp. 82–140.

80. IBARRA, O. H., AND KIM, C. E. Fast approximation algorithms for the knapsack and sum of subset problems. *Journal of the ACM 22*, 4 (1975), 463–468.

81. JENSEN, K., AND WIRTH, N. *Pascal. User Manual and Report*. Springer, Berlin, 1974.

82. JOHNSON, D. S. Approximation algorithms for combinatorial problems. In *Proceedings of the 5th Annual ACM Symposium on Theory of Computing* (1973), pp. 38–49.

83. JOHNSON, D. S. A brief history of NP-completeness, 1954-2012. *Documenta Mathematica* (2012).

84. KARP, R. M. Reducibility among combinatorial problems. In *Complexity of Computer Computations*, R. E. Miller and J. W. Thatcher, Eds. Plenum, New York, 1972, pp. 85–103.

85. KARP, R. M. On the computational complexity of combinatorial problems. *Networks 5*, 1 (1975), 45–68.

86. KARP, R. M., AND MILLER, R. E. Parallel program schemata. *Journal of Computer and System Sciences 3* (1969), 147–195.

87. KLEENE, S. C. *Introduction to Metamathematics*. Van Nostrand, New York, 1952.

88. KNUTH, D. E. *The Art of Computer Programming. Vol. 1: Fundamental Algorithms*. Addison-Wesley, 1968.

89. KNUTH, D. E. Semantics of context-free languages. *Mathematical Systems Theory 2*, 2 (1968), 127–145.

90. KNUTH, D. E. *The Art of Computer Programming, Vol. 2 : Seminumerical Algorithms*. Series in Computer Science and Information Processing. Addison-Wesley, Reading, 1969.

91. KNUTH, D. E. *The Art of Computer Programming, Vol. 3: Sorting and Searching*. Reading. Addison-Wesley, Massachusetts, 1973.

92. KNUTH, D. E., AND PARDO, L. T. The early development of programming languages. *A history of computing in the twentieth century* (1980), 197–273.

93. KOLMOGOROV, A. N. Three approaches to the quantitative definition of information. *Problems of Information and Transmission (Problemy Peredachi Informatsii) 1*, 1 (1965), 1–7.

94. KOWALSKI, R. Predicate logic as a programming language. In *IFIP Congress 74* (1974), pp. 556–574.

95. KOWALSKI, R. *Logic for Problem-Solving*. North-Holland Publishing, Amsterdam, The Netherlands, 1986.

96. KOWALSKI, R. The early years of logic programming. *Communications of the ACM 31*, 1 (Jan. 1988), 38–43.

97. KUNG, H. Let's design algorithms for VLSI systems. In *Proceedings of the Caltech Conference On Very Large Scale Integration* (1979), pp. 65–90.

98. KUNG, H. T. Why systolic architectures? *Computer 15*, 1 (Jan. 1982), 37–46.

99. KUNG, H. T., AND LEISERSON, C. E. Algorithms for VLSI processor arrays. In *Introduction to VLSI systems*. Reading, MA: Addison-Wesley, 1980, pp. 271–292.

100. LAWLER, E. L., LENSTRA, J. K., RINNOOY KAN, A. H. G., AND SHMOYS, D. B., Eds. *The Traveling Salesman Problem: a Guided Tour of Combinatorial Optimization*. J. Wiley & Sons, 1985.

101. LEVIN, L. A. Universal sequential search problems. *PINFTRANS: Problems of Information Transmission (translated from Problemy Peredachi Informatsii (Russian)) 9* (1973), 115–116.

102. LEWIS, P. M., STEARNS, R. E., AND HARTMANIS, J. Memory bounds for recognition of context-free and context-sensitive languages. In *Proceedings of the 6th Annual Symposium on Switching Circuit Theory and Logical Design* (1965), pp. 191–202.

103. LINDENMAYER, A. Mathematical models for cellular interactions in development, I & II. *Journal of Theoretical Biology 18* (1968), 280–315.

104. LIPTON, R. J., AND TARJAN, R. E. A separator theorem for planar graphs. *SIAM Journal on Applied Mathematics 36*, 2 (1979), 177–189.

105. LISKOV, B., AND ZILLES, S. Programming with abstract data types. In *Proceedings of the ACM SIGPLAN Symposium on Very High Level Languages* (1974), pp. 50–59.

106. LISKOV, B. H., AND ZILLES, S. N. Specification techniques for data abstractions. *IEEE Transactions on Software Engineering*, 1 (1975), 7–19.

107. LUCCIO, F. *Strutture linguaggi sintassi: una introduzione*. Bollati Boringhieri, 1987.

108. LUCCIO, F., Ed. *L'informatica. Lo sviluppo economico, tecnologico e scientifico in Italia*. Edifir, 2007.

109. LUCE, R. D., BUSH, R. R., AND EUGENE, G. E. *Handbook of Mathematical Psychology*. J. Wiley and Sons, New York, London, Sydney, 1963.

110. LUCKHAM, D. C., PARK, D. M. R., AND PATERSON, M. S. On formalised computer programs. *Journal of Computer and System Sciences 4*, 3 (1970), 220–249.

111. MAIER, D. Minimum covers in relational database model. *Journal of the ACM 27*, 4 (1980), 664–674.

112. MANNA, Z. *The Mathematical Theory of Computation*. McGraw-Hill, 1974.

113. MANNA, Z., AND VUILLEMIN, J. Fixpoint approach to the theory of computation. *Communications of the ACM 15*, 7 (1972), 528–536.

114. MCCARTHY, J. Recursive functions of symbolic expressions and their computation by machine, Part I. *Communications of the ACM 3*, 4 (1960), 184–195.

115. MCCARTHY, J. History of LISP. In *History of Programming Languages I* (1978), ACM, pp. 173–185.

116. MEAD, C., AND CONWAY, L. *Introduction to VLSI Systems*. Addison-Wesley Longman Publishing, Boston, MA, USA, 1979.

117. MEYER, A. R., AND STOCKMEYER, L. J. The equivalence problem for regular expressions with squaring requires exponential space. In *Proceedings of the 13th Annual Symposium on Switching and Automata Theory* (1972), pp. 125–129.

118. MILLER, R. E. Some undecidability results for parallel program schemata. *SIAM Journal on Computing 1*, 1 (1972), 119–129.

119. MILNER, R. Implementation and applications of Scott's logic for computable functions. *ACM SIGPLAN Notices 7*, 1 (Jan. 1972), 1–6.

120. MINSKY, M. *Computation: Finite and Infinite Machines*. Prentice-Hall, Englewood Cliffs, NJ, 1967.

121. MINSKY, M., AND PAPERT, S. *Perceptrons*. MIT Press, Cambridge, MA, 1969.

122. MOSSES, P. D. Denotational semantics. In *Handbook of Theoretical Computer Science*, J. van Leeuwen, Ed., vol. B: Formal Models and Semantics. The MIT Press, New York, NY, 1990, ch. 11, pp. 577–631.

123. MULLER, D. E., AND PREPARATA, F. P. Finding the intersection of two convex polyhedra. *Theoretical Computer Science 7*, 2 (1978), 217–236.

124. NASTASI, P. Picone, il calcolo automatico e FINAC, una storia lunga 30 anni. In *50 anni di informatica in Italia*. Università Bocconi, 2005.

125. NATIONAL ACADEMY OF ENGINEERING. *Memorial tributes*, vol. 12. National Academies Press, Washington, D.C., 2008.

126. NAUR, P. The design of the GIER ALGOL compiler. *Annual Review in Automatic Programming 4* (1964), 49–85.

127. NIVAT, M. The true story of TCS. *Theoretical Computer Science, 591* (2015), 1–2.

128. ORPONEN, P., AND MANNILA, H. On approximation preserving reductions: Complete problems and robust measures (revised version). Tech. Rep., Department of Computer Science, University of Helsinki, 1990.

129. PAPADIMITRIOU, C., AND YANNAKAKIS, M. Optimization, approximation, and complexity classes. In *Proceedings of the 20th Annual ACM Symposium on Theory of Computing* (1988), pp. 229–234.

130. PAZ, A. *Introduction to Probabilistic Automata*. Academic Press, 1971.

131. PAZ, A., AND MORAN, S. Non-deterministic polynomial optimization problems and their approximation. In *Analysis and Design of Algorithms in Combinatorial Optimization*. Springer, 1981, pp. 1–35.

132. PIPPENGER, N. On simultaneous resource bounds. In *Proceedings of the 20th Annual Symposium on Foundations of Computer Science* (1979), pp. 307–311.

133. PREPARATA, F. P., AND HONG, S. J. Convex hulls of finite sets of points in two and three dimensions. *Communications of the ACM 20*, 2 (1977), 87–93.

134. PREPARATA, F. P., AND SHAMOS, M. I. *Computational Geometry: An Introduction*. Springer, 1985.

135. PREPARATA, F. P., AND VUILLEMIN, J. The cube-connected cycles: A versatile network for parallel computation. *Communications of the ACM 24*, 5 (1981), 300–309.

136. PUDLÁK, P., AND SPRINGSTEEL, F. N. Complexity in mechanized hypothesis formation. *Theoretical Computer Science 8*, 2 (1979), 203–225.

137. RABIN, M. O. Degree of difficulty of computing a function and a partial ordering of recursive sets. Tech. Rep. 2, Hebrew University, 1960.

138. RABIN, M. O. Theoretical impediments to artificial intelligence. In *IFIP Congress 74* (1974), pp. 615–619.

139. RABIN, M. O., AND SCOTT, D. Finite automata and their decision problems. *IBM Journal of Research and Development 3*, 2 (1959), 114–125.

140. ROGERS, H. *Theory of Recursive Functions and Effective Computability*. McGraw-Hill, New York, 1967.

141. ROZENBERG, G., AND SALOMAA, A. ICALP, EATCS and Maurice Nivat. *Theoretical Computer Science 281*, 1 (2002), 25–30.

142. RUSTIN, R., Ed. *Formal Semantics of Programming Languages*. Prentice-Hall, Englewood Cliffs, NJ, 1972.

143. RYDEHEARD, D. E., AND SANNELLA, D. T. A collection of papers and memoirs celebrating the contribution of Rod Burstall to advances in computer science. *Formal Aspects of Computing 13*, 3 (2002), 187–193.

144. SAHNI, S., AND GONZALEZ, T. P-complete approximation problems. *Journal of the ACM 23*, 3 (1976), 555–565.

145. SALTON, G. *Automatic Information Organization and Retrieval.* McGraw-Hill, New York, US, 1968.

146. SAMMET, J. E. *Programming Languages: History and Fundamentals.* Prentice-Hall, Englewood Cliffs, NJ, 1969.

147. SCOTT, D. S. Outline of a mathematical theory of computation. In *Proceedings of the Fourth Annual Princeton Conference on Information Sciences and Systems* (1970), pp. 169–176.

148. SCOTT, D. S. The lattice of flow diagrams. In *Symposium on Semantics of Algorithmic Languages* (1971), E. Engeler, Ed., vol. 188 of *Lecture Notes in Mathematics*, Springer.

149. SCOTT, D. S. A type-theoretical alternative to ISWIM, CUCH, OWHY. *Theoretical Compututer Science 121*, 1-2 (1993), 411–440.

150. SCOTT, D. S., AND STRACHEY, C. Toward a mathematical semantics for computer languages. In *Proceedings of the Symposium on Computers and Automata* (1971), Polytechnic Institute of Brooklyn Press, pp. 19–46.

151. SHASHA, D., AND LAZERE, C. *Out of Their Minds: the Lives and Discoveries of 15 Great Computer Scientists.* Springer, 1998.

152. SLEATOR, D. D., AND TARJAN, R. E. A data structure for dynamic trees. *Journal of Computer and System Sciences 26*, 3 (1983), 362–391.

153. SOLOMONOFF, R. A preliminary report on a general theory of inductive inference. *Revision of Report 131* (1960).

154. SOLOMONOFF, R. J. A formal theory of inductive inference. *Information and Control 7* (1964), 1–22, 224–254.

155. STEARNS, R. E. Juris Hartmanis: the beginnings of computational complexity. In *Complexity Theory Retrospective.* Springer, 1990, pp. 5–18.

156. STEEL, T. B. *Formal Language Description Languages for Computer Programming.* North-Holland, New York, 1966.

157. STOCKMEYER, L. J. The polynomial-time hierarchy. *Theoretical Computer Science 3*, 1 (1976), 1–22.

158. STRACHEY, C. Towards a formal semantics. In *IFIP TC2 Working Conference on Formal Language Description Languages for Computer Programming* (Amsterdam, 1966), North-Holland, pp. 198–220.

159. TARJAN, R. Depth-first search and linear graph algorithms. *SIAM Journal on Computing 1*, 2 (1972), 146–160.

160. TARJAN, R. E. Efficiency of a good but not linear set union algorithm. *Journal of the ACM 22*, 2 (1975), 215–225.

161. TURING, A. M. On computable numbers, with an application to the Entscheidungsproblem. *Proceedings of the London Mathematical Society 2*, 42 (1936), 230–265.

162. ULLMAN, J. D. The U. R. strikes back. In *Proceedings of the 1st ACM SIGACT-SIGMOD Symposium on Principles of Database Systems* (1982), pp. 10–22.

163. VALIANT, L. G. Completeness classes in algebra. In *Proceedings of the 11th Annual ACM Symposium on Theory of Computing* (1979), pp. 249–261.

164. VALIANT, L. G. The complexity of computing the permanent. *Theoretical Computer Science 8*, 2 (1979), 189–201.

165. VALIANT, L. G. The complexity of enumeration and reliability problems. *SIAM Journal on Computing 8*, 3 (1979), 410–421.

166. VALIANT, L. G., AND BREBNER, G. J. Universal schemes for parallel communication. In *Proceedings of the 13th Annual ACM Symposium on Theory of Computing* (1981), pp. 263–277.

167. WAGNER, K., AND WECHSUNG, G. *Computational Complexity.* Springer, 1989.

168. WEISSMAN, C. *Lisp 1.5 Primer.* Dickenson Publ., Belmont, CA, 1967.

169. YAO, A. C. Protocols for secure computations. In *Proceedings of the 23rd Annual Symposium on Foundations of Computer Science* (1982), pp. 160–164.

170. YAO, A. C. Theory and application of trapdoor functions. In *Proceedings of the 23rd Annual Symposium on Foundations of Computer Science* (1982), pp. 80–91.

171. YOUNG, P. Juris Hartmanis: Fundamental contributions to isomorphism problems. In *Complexity Theory Retrospective*. Springer, 1990, pp. 28–58.

172. ZADEH, L. A. Fuzzy sets. *Information and Control 8*, 3 (1965), 338–353.